Raising Rabbits

Beginners Guide to Raising Healthy and Happy Rabbits

By: Irene Mills

Table of Contents

Introduction

Rabbits have been domesticated and raised for meat and also as cuddly, adorable pets. Their genetic ancestors are still roaming in wild places, and just as wild rabbits hold a place in a healthy ecosystem, domestic rabbits can be homestead or gardening partners. As prey animals, they think differently than dogs and cats, even though we might cuddle them just as we do other pets.

There are over fifty species of domestic rabbits around the world. Most of the domestic rabbits are in the ten-pound range. Some dwarf rabbits weigh as little as two to three pounds. Some wild rabbits are the same size as a small child! Of the thirty different species of wild rabbits, many live in several kinds of environments. From the desert to icy and snowy environments, rabbits can adapt and survive almost anywhere. Wild rabbits are critical to the food chain for this planet.

This book, *Raising Rabbits*, will be full of information and a guide to gaining the most profit from your new project. With care, attention, and good hygiene, rabbits are easy to raise and are relatively inexpensive.

Part One introduces you to rabbits and their world. For any owner, it is helpful to understand a rabbit's outlook and how their senses work. We will also discuss the wild rabbits who formed the basis of our domestic breeds and how they fit into the environment.

Part Two is all about getting started. Choosing a rabbit breed means that you must know *why* you want rabbits. We will discuss various functions of rabbits, including pets, show, agility, and meat. When we discuss each function, the recommended beginner breeds will be listed and described. You will read about the best hutches and see examples of equipment for food and water as well as handling and grooming. Using this book as a guide, you can be prepared and ready to welcome your new rabbits.

Part Three will discuss the maintenance required to keep your rabbits healthy and happy. Rabbits need to be kept healthy physically and mentally. This section will show you how to keep them this way and provide fun options for playtime with your rabbits. We will end with a section about the fascinating subject of rabbit behavior.

Why do they thump? What are they saying with their ear movements and body language? Did I just hear a vocal sound from my rabbit? We will answer these questions and more in the final section.

Please note, this book is written for an individual or family who wants to have rabbits for their own purposes, whether it be as a family pet or a small family homestead. Raising rabbits on a larger scale for profit is not part of the scope of this book.

Let's get started by learning about the rabbit's world and how they think with their senses.

Part One: Rabbits and Their World

Fun Facts

- Some species of wild rabbits can reach speeds of up to 50 mph.[1] They can keep up high speeds for relatively long distances as well. Domestic rabbits will not reach this kind of speed, so you won't have a racing bullet in your backyard.

- Rabbits love to play, and they need stimulation so they are not bored. They will enjoy a good "chew and toss" session with toilet paper rolls or non-dyed cardboard cut into pieces. If you're using cardboard boxes, include some corners and angles to make it different and interesting for them. Once the bunny has played with it, it will compost or recycle easily.

- You may not find this to be a "fun fact"—it's more of a "weird fact" with an "eww" factor. Rabbits eat their excrement, often as it comes out of their body. *Coprophagia* is normal and natural; all rabbits do this. They are trying to get nutrition from the food that passed through their digestive system. When you get a rabbit, don't be alarmed if you see them curled in on themselves eating their poop as it comes out, it's normal.

Rabbit Senses

Whether you have just one rabbit as a pet or are raising a colony of rabbits on a homestead for food, it is beneficial to understand your rabbits' vision, hearing, and smell functions to understand their needs and behavior.

This section will introduce the vision, hearing, and olfactory senses of rabbits and provide excellent resources if you decide to learn more.

Vision

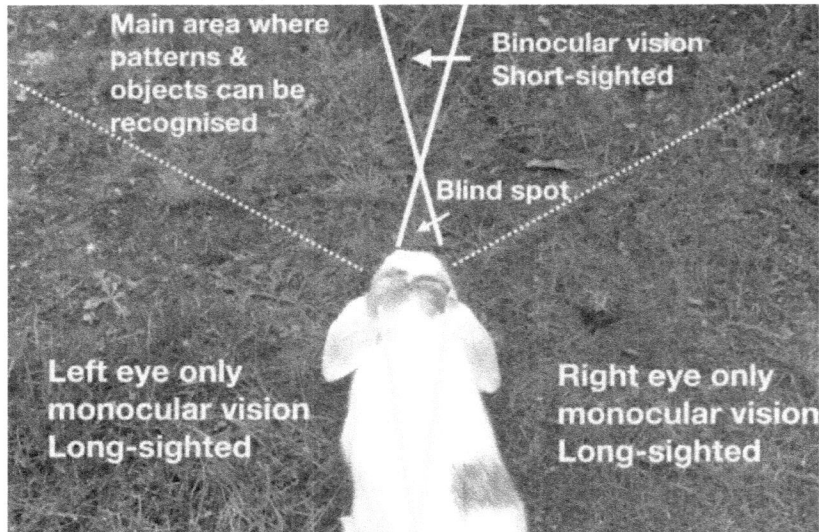

"How Rabbits See," newrabbitowner.com[2]

Rabbits do not have vision as sharp as ours, but they see better than us in low light, and their peripheral vision is fantastic—almost 360 degrees around. The rabbit needs to see an attacker coming from any direction, so having eyes on the side enables them to see a wraparound view of the world and keep them safer.

As prey animals, rabbits have evolved to be able to see at far distances. Interestingly, as you can see in the above diagram, they also see very well in short distances (presumably for foraging). This "dual-mode" vision makes them *both* long- and short-sighted. They will rely on their sense of smell for objects right in front of their nose.

Seeing Color

Rabbits do see color, but they do not easily distinguish between greens and reds and blues and greens.

In the color test below, bunnies would only be able to make out the number 2. Most humans see the number 42. Some people who are color-blind only see the number 4 or the number 2.

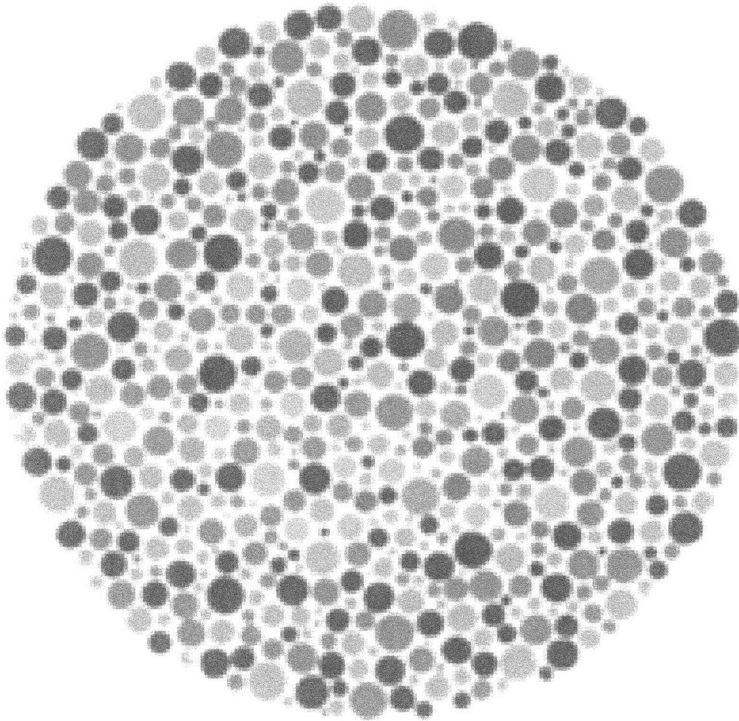

Color Blind Test, nymag.com[3]

Rabbits (compared to most humans) are slightly color-blind (protanopic or dichromatic).

But there's another difference between rabbits and humans.

Location of *Rods and Cones in Rabbits* – and why it's important.

Rods and cones both pick up color, but cones are more tuned explicitly to color differentiation. Your bunny will not see things the same way because their size and needs are different.

As humans, we are used to seeing detail in color when we look straight ahead. This focus is because our cones are clustered in an indent right in the center. Our rods that see better in dim light are denser in our peripheral vision.

On the other hand, rabbits have their color-sensitive cones in the "visual streak" across the retina. This structure is what helps them have more detail in their peripheral vision. If we humans want to look at something in detail, we will focus on it straight on. A rabbit will ensure that they are using their peripheral vision to get more detail. The band of cones in the visual streak is part of their adaptation for survival.

Added to the visual streak of cones, a blue streak lays in a crescent shape at the bottom of the retina. The placement of the blue streak enables the rabbit to see attacks coming from aerial predators such as owls and hawks.

When you observe your rabbits, you can now understand that they will be utilizing their visual streak to scan for predators on the ground and the blue streak to keep a watch on the sky.

Night Vision

Rabbits are *crepuscular,* which means they are most active during dawn and dusk.[4] As a result, their vision is better than ours in low light, but like us, they are not adapted to seeing in complete darkness like nocturnal animals.[5]

Do Rabbits Blink?

Yes, they do—but not as often as we do. This is sometimes a concern for new rabbit owners. Rabbits have more moisture in their eyelids and only blink approximately once every five minutes.

Hearing: Rabbits and "Those Ears!"

Structure

As we've discussed, the rabbit species has eyesight that evolved to pick up movement. This enables them to track predators visually. Their hearing, with their huge "satellite dishes," will pick up things they can't see or before they are visible.

This section will discuss the basics of the ears, their varied functions (not just hearing), and possible conditions and complications.

The Outer Ear

The outer ear is comprised of two parts:
- the visible pinna,
- and the outer ear canal.

The *pinna* serves your rabbit in several ways:
- Each ear can rotate nearly 270 degrees. This helps the rabbit zoom in on the location of a sound and keep track of two different sounds simultaneously.[6]
- With a lot of blood supply and a large surface area, the *pinna* also functions as a temperature regulator. In the heat, the ears will help the rabbit cool off.
- You can help your rabbits feel more comfortable and avoid heatstroke by applying a cool (but not cold), damp cloth to their ears. Compresses work for bringing down a fever as well.
- If you live in a place with cold winters, the blood vessels in the *pinna* will constrict. Talk to your vet about whether there is any threat of frostbite to your rabbit's ears. How much risk your rabbit has of frostbite will depend on the lowest temps, the length of time of those lowest temperatures, and the breed of rabbit.
- Rabbits use the *pinna* part of their ears to communicate almost constantly. More details of rabbit communication using their ears are in the next section.

The outer ear canal leads from the outside of the ear to the eardrum.
- Rabbits can get wax buildup and are prone to ear infections.

- For the most part, rabbits will clean their ears of wax, dirt, and plant debris, but it is good to know how to do it yourself. Ask your vet to show your how to clean your rabbit's ears.
- Rabbits are prone to ear mites that will concentrate in the outer ear canal and the *pinna.* Your vet can prescribe ear mite treatment.

The Middle Ear

The middle ear is deeper; it is beyond the eardrum. Bacteria can get down into the middle ear, causing pus and pain to your rabbit. If you see pus, get to your vet as soon as possible.

The Inner Ear

The inner ear includes more parts that provide detail of pitch, volume, and precise detail of the sound (such as the difference between the soft crunch of a leaf under a bobcat's foot versus the snap of a twig under a deer hoof).

The inner ear also includes the *labyrinth,* which does not function for hearing but for balance, movement, and proprioception.[7]

A rabbit owner should be aware of the symptoms of labyrinth damage that can result from either disease or injuries.
- A "head tilt characterizes a damaged labyrinth." The eyes will be moving side to side, and the head will tilt toward the side of the damaged ear. *"We see this as rolling or circling in rabbits with ear infections or neurological diseases such as Encephalitozoon cuniculi."*[8]
- Sometimes the labyrinth will be under pressure because of a middle ear infection.

Below is a diagram of the anatomy of a rabbit's ear.

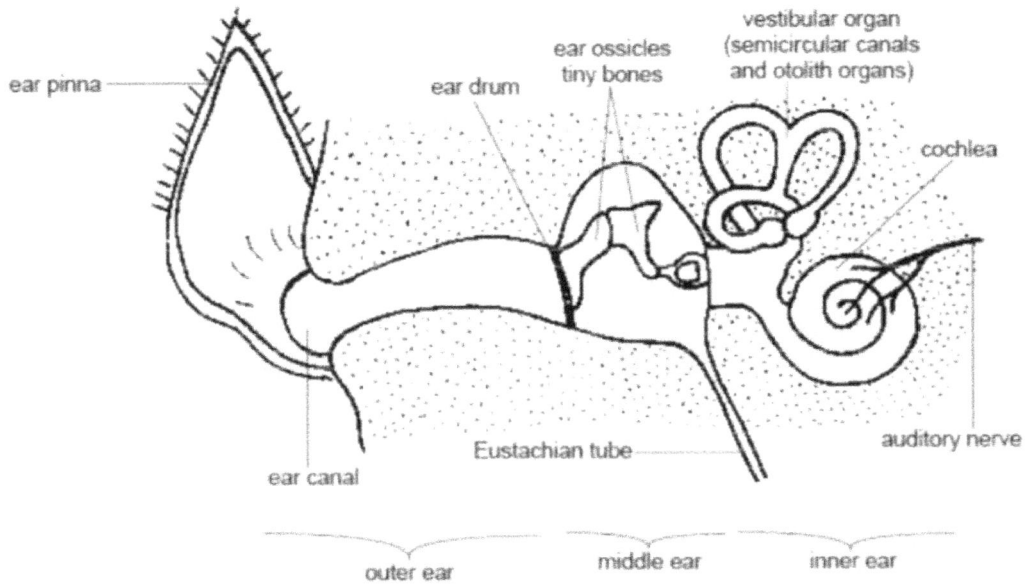

Sense of Smell

Rabbits have a keen sense of smell. They have 100 million receptors in their nose and the ability to move a scent around in their nose to help identify it. When you see a rabbit moving its nose up and down, it is assessing the details of a scent. This is called "nose blinking."[10]

Your rabbit's sense of smell will be 20X "louder" than yours. We humans have five to six million scent receptors. *A rabbit has up to a hundred million!*[11] Cats have about eight million.[12] Rabbits have developed this keen sense to find the best food, sense predators, find each other, and read "messages" left by their colony and other animals. Rabbits rub and leave a scent to communicate ownership of a person, object, or territory. A wild rabbit can even smell underground food.[13] Rabbits use their sense of smell to fill in details that their vision does not give them.

In addition to a hundred million scent receptors, rabbits have the "Jacobsen's organ." This spot on the roof of the mouth of a lot of mammals detects scent. While other animals open their mouth to use their Jacobsen's organ, rabbits do not open their mouth because they have a split lip that allows air to flow.

Pet rabbits retain the sharp senses of their wild relatives. You may notice that your rabbit's nose is usually on the move. When a rabbit's nose is twitching, it is working hard to zero in on a scent, and they are seeking more information.

While your pet rabbit may not have the predatory pressure that a wild rabbit lives with, they will be using that sense of smell to confirm and tag who is friend and foe. If you bond with your rabbit early on, your smell will be a comfort to them. If someone in your household has grabbed, chased, or frightened the rabbit, their sense of smell will tag that person as unsafe. That goes for house animals as well, like a dog or cat. The sense of smell is deeply primal and is linked to a mammal's emotional systems. Your rabbit will label "friend" or "foe" and then identify that person or animal as such in the future.

Predators

If you have a dog or cat in the house with a rabbit, you will need to manage and supervise carefully. *If* you can train the dog or cat to not chase or harass the rabbit, this *might* work. The thing is that dogs and cats are natural predators; the movement and smell of a rabbit can get to their primal instincts. Also, I've known people who thought their dog was trained not to chase small livestock such as goats, then one day their chase instinct goes off and they commit carnage.

If your rabbit hutch is in a house with an animal that they perceive to be a threat, the rabbit will be constantly stressed. Long-term stress can end up undermining their overall health. It can also make colony dynamics strange which can result in aggression and behavioral problems.

Unpleasant or Harmful Smells

When rabbits smell things they do not like, it will stress them. The sensitivity that serves the rabbit's survival means that unpleasant scents are a source of stress and sometimes a hazard to their health. Keep your rabbit away from the following things:

Chemical scents found in
- candles,
- cleaning supplies,
- laundry soaps,
- air fresheners, and
- perfumes and cologne.

These items are likely to be irritating and even dangerous to your rabbits.

Keep chemical scents away from their enclosures, hutch, and your home if they are kept inside. If you use strongly scented laundry soap, consider having a shirt to put on for handling your rabbits washed in fragrance-free soap so they don't have to endure the onslaught of chemical fragrance when they are next to you.

Chemical fragrance is essential to manage because it can irritate the tissue in the nasal passages and result in an upper respiratory infection.

According to Lou Carter, there are three other common scents to avoid because they are repellent to rabbits.

Lavender

Humans often find this plant to be calming. For some reason, rabbits hate it, and will even become agitated when they are exposed to it. Keep the natural plant and any lavender-scented products away from your rabbit, and they will thank you.

Onions and garlic

Rabbits hate these smells. If you let your rabbits roam the house and you want to keep them out of a closet, all three of the above are good choices to use as repellents. Just make sure the scent isn't powerful to your sense of smell, and that you can contain it away from their hutch, so they aren't tortured by it in their own bedroom.

Spicy scents

Rabbits do not like the smell of chilis, mustards, and other strong spices. It is possible that the rabbit can feel the hot spices in those sensitive noses. Take care to close spice jars in your kitchen when you use them.

Providing an area that is free of anxiety-producing smells is extremely important for your rabbit. Their sense is so keen; one writer compares the experience of a rabbit in a home with smells (e.g., a dog, a cat, and air fresheners) to living with a heavy metal band that plays loudly 24/7.[14]

Conclusion of Rabbit's Senses

This section has provided you with a basic understanding of the rabbit's vision, hearing, and smell. Watch for the signs mentioned, such as discharge, mites, or the head tilt associated with inner ear damage. It is also advised to talk to your vet about any specific vulnerabilities of the breed of rabbit you have and protocols for assessment and treatment.

How Rabbits Move

The baseline movement for a rabbit is a hop. They can cover a lot of ground quickly by hopping. They will often walk when they are moving in a small space, grazing from plant to plant. They will also walk if they are taking care to explore a new area. You'll see them walking as they "nose blink" and have their ears straight up and wide open, listening.

While rabbits use a walk for sensing purposes, it could also indicate an injury if your rabbit is walking instead of hopping. It can also be a sign of anxiety—is this rabbit being bullied, or is it afraid of other rabbits? Ensure your rabbit is healthy (illness or injury can make a rabbit not want to hop) and that all is well in the colony dynamics.

Rabbits Who Would Rather Walk

Sometimes, a constantly anxious rabbit will walk because they are not comfortable or confident in their territory or with their colony. Observe the behavior dynamics between rabbits and intervene, if necessary, as described in the Rabbit Behavior section. Also, continue to give your nervous rabbit positive reinforcement to build their confidence in their territory.

Besides anxiety, not hopping can also signify the rabbit feeling weakness in their hind legs. Contact your vet for examination; sometimes, not hopping is a sign of a health issue that does not present in limping or dragging.

Rabbits in the Wild

Habits

Rabbits are *crepuscular*; they are most active at dawn and dusk. It is interesting that some of their most threatening predators also hunt at that time. According to livescience.com, the low light makes them feel less vulnerable to being seen.[15]

Wild Rabbit Species

Familiar domestic rabbits inhabit our hutches, but other interesting wild rabbits and hares are living around the world. Here is a list of wild rabbits who are the most relevant to homesteaders and rabbit owners in the US.

Eastern Cottontail
welcomewildlife.com[16]

Cottontails

Cottontails are among the most common wild rabbits in North America. They reside throughout almost all of North America but are more common in America's central and midwestern parts. Many can also be found in the northern and central parts of South America. Cottontails are mostly gray and brown in appearance and are named after their classic white tail.

There are eight different breeds of cottontails, each residing in different environments and regions of America. Among these are the Desert Cottontail, Eastern Cottontail, New England Cottontail, the Brush Rabbit, and Swamp Rabbit.

Jackrabbits

Jackrabbit, Wikimedia Commons[17]

Jackrabbits are some of the most interesting-looking hares. With tall limbs, a lanky body, large feet, and long ears, they are well equipped to run extremely fast to escape predators. In fact, they can run at 37 mph. Black-tailed Jackrabbits can run within a few hours after birth. Their cousins, the White-tailed Jackrabbits, are mostly nocturnal, which they adapted to escape birds of prey during the day.

Another outstanding breed is the Antelope Jackrabbit. They have large ears that enable them to hear predators approaching from long distances and keep them from overheating during the hot summers of Arizona and Mexico.

Snowshoe Hare

The snowshoe hare is a common hare that is distributed across the snowy reaches of Alaska and Canada. Most have white fur and shorter ears because of adapting to their environment. Their white fur allows them to blend in with the snow; Snowshoe Hares turn brown during the summer.

Snowshoe Hare in Summer   Snowshoe Hare in Winter

"Snowshoe Hare," Eleur.com[18]

Two other species of hares live in the far north:

- The Arctic Hare, *Lepus arcturus*, lives in the frozen tundra of Canada north of Alaska
- The Alaskan Hare, *Lepus othus,* lives in the coastal regions of Alaska, reaching not much farther north than Nome.

These hares that live in very cold climates have much shorter ears than are typical for hares to prevent heat loss.

Additionally, their breeding season is very short, resulting in only one litter per year. As a result, their litters are much larger than the average wild rabbit's litter.

Arctic Hare, Wikimedia Commons[19]

Pygmy Rabbits

US Fish & Wildlife Service "Pygmy Rabbit"[20]

Pygmy Rabbits, *Brachylagus idahoensis*, are native to the northwestern areas of the United States. They live in a sagebrush/desert ecosystem. At one point, this tiny rabbit breed was endangered due to expanding farmland and the destruction of their habitat. However, their species is starting to recover due to the combined efforts of biologists, zoos, and universities. At the time of this writing, the conservation status on Wikipedia indicated "least concern."[21]

European Wild Rabbits

Wikimedia Commons – "European Rabbit Lake District UK"[22]

A very interesting fact for rabbit breeders and for those who have pet rabbits is that all pet and domesticated rabbits descend from the European Rabbit.[23] This is the rabbit we see at every rabbit show, our 4-H rabbits, rabbits for meat and fur, and our loving household pet rabbits.

In the wild, the European Rabbit is brown in appearance, with a short tail and long ears. Their lives center around eating and breeding. They graze constantly, tearing up and foraging any greenery within their sight. When they are not grazing, they are breeding, producing at least 30 to 40 offspring per year, and teenagers can become fertile in only 3 to 4 months after birth.

As a result, the European Rabbit is by far the most common rabbit of all time and overpopulates many areas. It was first domesticated in ancient Rome, and since the Middle Ages, this rabbit has been spread all over the earth. In places such as Australia, the European Rabbit runs rampant, invading and devastating the ecosystem. Ironically though, its population is declining in its native lands.

Wild Rabbits in Australia

Australian Rabbits are actually European Rabbits. The European Rabbit has taken up an unwelcome residence in Australia. Ever since the early 1800s when English landowners unwittingly released rabbits into the wild to hunt them for meat, the European Rabbit, *Oryctolagus cuniculus*, has been multiplying nonstop.

"Pest" and "invasive" are understatements when it comes to rabbits in Australia. With no natural predators down under, they have been multiplying exponentially for 150 years. These rabbits consume vegetation nonstop and destroy crops and fields intended for cattle grazing, resulting in financial losses in the millions of dollars. The over-foraging also affects the other species around them, as they eat most of the resources that other species need to survive.

As the feral rabbit population topped 400 million, the rabbit disease known as myxomatosis was released into the environment with excellent results. Possibly 99% of feral Australian rabbits died. But that other 1%? They already had, or developed thanks to excellent immune systems, natural immunity to the disease and went right on about their business of multiplying.

After that, another round of biological warfare was tried against feral rabbits, the lethal viral disease known as rabbit hemorrhagic disease, or calicivirus. The results were identical. A very small population survived, and these immune rabbits are now multiplying at the same exponential rate as before.

It is estimated today that over 100 million feral rabbits still plague the Australian countryside, and the count continues to rise. Scientists continue trying to figure out what to do about the feral rabbits of Australia.

Rabbits and the Environment

Throughout history, rabbits had their place in the ecosystem, and nature held a balance. With the destruction of habitats, there are fewer large predators, and wild rabbits have become a destructive force in the environment.[24]

Keeping rabbits as pets can be a very environmentally-friendly choice. Since they are herbivores, their carbon footprint is lower than other pet choices.

How Rabbits Help the Environment

As we have mentioned, wild rabbits have a role in maintaining a balanced and healthy ecosystem.
- They maintain a balance in the vegetation.
- They nourish the soil.
- They provide food for carnivores and omnivores. They do not hibernate in the winter, so they offer available nutrition year-round.

Nature does its beautiful dance of balance as the rabbits maintain the vegetation, nourish the soil, and provide a food source while the predators keep the population down so that the rabbits don't become a destructive force.

Sadly, the late 20th and 21st centuries have seen that balance lost.

How Rabbits Cause Destruction in the Environment – A Warning to Rabbit Owners

The combination of loss of habitat through climate change (e.g., severe, long-lasting drought) and development has pushed rabbits into more density as well as lessened the number of predators.[25]

Another issue that is important for a rabbit owner to understand is that rabbits introduced to a new habitat become an invasive species. It does not take long for them to overpopulate. In some places where this has happened, the introduced rabbits end up competing with the native wild rabbits for food.[26]

Yet another consequence of introducing non-native species is that the ecosystem can get out of balance by attracting feral cats and too many foxes to a small space. This results in other native mammals becoming *"at risk of endangerment and extinction."*[27]

Lastly, non-native rabbits released into the wild are a pest for gardeners. When there is balance with a wild rabbit population and their natural predators, a gardener must protect their gardens, but they are not overtaken by rabbits. In suburban areas, the rabbits released do not have enough predators to keep their numbers in check, and they become a serious problem.

How Climate Change is Affecting Rabbits

Climate change is affecting wildlife everywhere on the planet, and rabbits are no exception. They are having to adjust to rising temperatures as well as loss of habitat due to natural disasters such as fires and floods.

Scientists expect that two-thirds of all species of rabbit will be forced to migrate out of their current territory by 2100. They may migrate to cooler climates or relocate to higher elevations in mountainous areas.[28]

Part Two:
Getting Started

Choosing a Breed

Here we'll list breeds under sections based on their function and purpose in your life. As of 2017, there were at least 305 breeds of domestic rabbits in 70 countries around the world.[29] The American Rabbit Breeders Association registers more than 50 breeds. The breeds you will read about here are chosen as examples because they are the most popular, accessible, and highly recommended for the specific function where they are listed.

Clearly, there are pet rabbits that you might decide to show and fur rabbits that may also be used for meat. The 4-H organization is all about showing in one way or another, so you will find that many rabbits can be a part of your life as multi-functional.

We will discuss owning rabbits for the following functions:
- Pets
- Show
- 4-H
- Agility
- Meat and fur
- Homestead and gardening partners

Rabbits as Pets

Having a rabbit as a pet is as big of a commitment as having a cat or a dog. They are adorable, and many famous rabbits such as Thumper, Peter Rabbit, and Bugs Bunny have made having a pet rabbit almost irresistible. But before your enthusiasm takes you to a bunny, make sure that you are truly committed to the care needed for a rabbit. There are two schools of thought for how to house a pet rabbit. Some people swear by having a rabbit in a cage outside, and other rabbit owners gasp at the thought. The latter insist rabbits should be brought inside to be with the family.

One concept is clear. Rabbits are social animals, so unless you are going to be around your rabbit all the time, you should get two rabbits instead of one. Also, keep in mind your rabbit(s) are going to live eight to ten and sometimes twelve years if they are well cared for. Yearly visits to a knowledgeable rabbit vet will help ensure your rabbits are healthy and happy. Most of us are familiar with at least the basic care of cats and dogs, but fewer people know what to expect when they acquire a rabbit. This could explain why rabbits are the third most surrendered animals to shelters.

Rabbits of any breed are not recommended as pets for very young children. Little children want to rush up to rabbits and pick them up, and rabbits really do not enjoy being grabbed. They have a response like they have been grabbed by a predator. Wait until children are old enough to understand the proper treatment of animals before getting a rabbit.

Read the sections in this book about cleaning the hutch and the mental stimulation required for your rabbit. If you decide that a pet rabbit would fit into your lifestyle, you can set yourself up for great joy by choosing the right breed.

The following is a curated list of rabbit breeds that are especially suited as pets. They all have characteristics of being calm and sociable. Some are less active; others are energetic and will want to play. Some need more space than others. None of these rabbits are particularly vulnerable to health issues; they just need standard rabbit health care. None of these rabbits are particularly expensive. Here are our top recommendations for pet rabbit breeds based upon our experience and knowing other rabbit owners.

Mini-Rex

pets4-Homes.co.uk[30]

rabbitpedia.com[31]

SqueaksandNibbles[32]

Pets4-Homes.co.uk[33]

The Mini Rex is the #1 or #2 top entry at any ARBA Convention. It is not small enough to be categorized as a "Dwarf" rabbit, but is on the smaller side. Their size and fur make them a top choice as pet rabbits.

- **Weight**: 3.5–4.5 lbs.
- **Colors and Fur:** Their fur is extra plush and soft because the hairs on their fur are the same length, and they all stand up instead of lying flat. As you can see from the images, they can come in many colors, both solid and in combinations.
- **Temperament:** They are gentle and social. They love your company and like to play. They are also relatively energetic, so they *need* play and exercise to be

happy. These rabbits are also smart and trainable. Once you've earned their trust, you can train them to do tricks and use a litter box in the house.[34]

- **Special Considerations:** None.
- **Go here for more resources**[35]

American Fuzzy Lop

animalcorner.org [36]

American Fuzzy Lop Rabbit Club[37]

These lovely rabbits are sometimes described as looking comical and grumpy. They are far from grumpy; they are wonderful pets.

I've not kept American Fuzzy Lops as an adult, but my parents got one when I was 8 years old, and she was the *perfect* pet. She loved to be cuddled and stroked and was also playful. I learned a lot about caring for an animal's needs from that dear rabbit.

- **Weight:** 3–4 lbs.
- **Colors and Fur:** Their fur is dense and soft. There are many color combinations.
- **Temperament:** When handled from a young age and after gaining trust, they love to receive your attention and enjoy being cuddled.
- **Special Considerations:** Not tolerant of very high temperatures. Make sure that your lifestyle can accommodate regular handling and affection. Also, make sure there is enough stimulation inside the hutch as well as playtimes outside the hutch. If bored, they will start "acting out" by chewing on their cage or becoming depressed or irritable.[38]

Holland Lop

Petguide.com[39]

The Holland Lop is a dwarf rabbit and is recommended as being particularly suited to small apartments. You may also consider this breed if you have a small space where you expect to contain your bunny's adventures.

- **Weight:** 2–4 lbs.
- **Colors and Fur:** Their fur is short and dense and can come in a wide variety of color combinations.
- **Temperament:** Curious—needs stimulation. Also energetic and likes to play.
- **Special Considerations:** This is another rabbit that thrives on your attention, so make sure that your lifestyle can accommodate a lot of interaction and time for affection.

Mini-Lop

Petguide.com[40]

- **Weight:** 4.5–6 lbs.
- **Colors and Fur:** Medium-length fur that doesn't require a lot of grooming. They come in many colors and variations.
- **Temperament:** Described as the "cuddliest" of all the rabbit breeds. They love touch and affection. This rabbit is described as especially good for seniors as well as older children because they want to be a "lap rabbit."
- **Special Considerations:** Craves time outside.[41]
- **Lifespan:** 5–10 years

Havana

Littlefurrypets.com[42]

- **Weight:** 4.5–6.5 lbs.
- **Colors and Fur:** Low maintenance coat. They are beautiful, and known as "the mink of the rabbit family."[43] Colors recognized by the American Rabbit Breeders Association are "chocolate, black, blue, and broken colors."[44]
- **Temperament:** Less energetic than rabbits mentioned previously, they are a plus for an owner who would like a rabbit who is happy to "hang out" with them versus needing a lot of playtime.
- **Special Considerations:** Havanas need more "roaming" time than other breeds. This might suit an owner who works from home or has any lifestyle of being around the house a lot of hours during the day. They are less energetic than rabbits mentioned previously.[45]
- **Lifespan:** 5–8 years

Havanas are also wonderful as show rabbits. [46]

Polish

AmazingPetsForYou[47]

The Polish rabbit is not from Poland. The breed was developed in the 19th century in England. The word "polish" refers to the polish of its beautiful fur. It is a rabbit that serves as both a wonderful pet and a fine show animal.

- **Weight:** 3–3.5 lbs.
- **Colors and Fur:** Short fur that looks "polished." There are many recognized color varieties.
- **Temperament:** Loves to be around their humans. Affectionate, calm, and gentle, but also energetic.
- **Special Considerations:** This breed is energetic and needs playtime and exercise as well as cuddling.
- **Lifespan:** 5–6 years

Mini-Satin

Pets Nurturing[48]

- **Weight:** 3–4.5 lbs.
- **Colors and Fur:** The site in this endnote lists fourteen colors.[49] It is recommended that a slicker brush be used for this breed because it was developed especially for them. They only need brushing once or twice a week and a bit more when they shed their winter coat.[50]
- **Temperament:** Playful and love to interact with people they trust.
- **Special Considerations:** None.
- **Lifespan:** 5–8 years

Rabbits for Show

It is one thing to keep your pet rabbits happy and healthy. It is another to condition and train a rabbit for the purposes of showing. As with all animals, each breed will have its "breed standard" for measurements, proportions, and fur quality. For some breeds, there are also required poses specific to that breed. Show rabbits require specific care and training to be at their best when they compete.

Health Requirements for Show Rabbits
- No pinworms.
- No loose stools or diseases, such as coccidiosis or abscesses.
- No fur mites and ear mites (these are grounds for disqualification).
- No white snot. Don't show a rabbit that has *EVER* shown any chronic runny nose. Even if it's been a while since you last saw or heard any sneezing, the stress of leaving familiar surroundings and traveling to the show may precipitate a relapse. Putting sick rabbits on the show table is a waste of your money and a danger to every other rabbit. Any white snot or evidence of it such as matting on the front paws are disqualifications. Mats mean snuffles.

Rabbits Must Be Clean and Groomed
- No urine or fecal stains in the fur.
- Well-groomed with no loose fur or mats.
- Toenails clipped, so they cannot snag on the wire and rip out or lay a bloody laceration across the judge's hand or arm. A missing toenail is a disqualification if the judge cannot find evidence of it, even if it is still present but ripped to the skin and not grown out yet.

Rabbits Should Look Healthy; Well Nourished but Not Overweight
- Give all the fresh, clean water the rabbit can drink.
- Feed a fresh, 16–17% balanced feed, milled not more than two months prior.
  - *"Once your rabbit has achieved prime condition and coat, cut back slightly on the amount of feed you give the rabbit without starving the animal. This will maximize the length of time that he stays in condition and his coat stays in prime condition."*[51]
- Free-feed your growing juniors, but limit the rations of your senior (adult) rabbits. If there's feed in the feeder the next day, cut back on the pellets. If the rabbit dives

desperately at the feed like a drowning swimmer clawing for survival, then maybe it needs a bit more feed!
- Provide grass hay every day or very frequently to prevent fur blockage.
- Two weeks to a month before the show, increase the feed slightly or add a top dressing to the rabbit's rations. See Formula for Rabbit Fur Conditioning below. This formula is great for wool as well.

Rabbits Should Be Housed in a Well-Ventilated and Clean Environment

Lack of ventilation and sanitation are the causes of most common rabbit diseases. Just keeping your rabbit's hutch well maintained will make it much easier to ensure your rabbit is healthy.

Excellent ventilation helps keep ammonia odors at a minimum. Rabbits are susceptible to *pasteurellosis* if they are kept where the ammonia from their urine has become too strong. The ammonia burns the lungs of rabbits. Even if they are super hardy and don't get sick, they will not be in the best condition for showing if their bodies are trying to heal their lungs constantly.

Some rabbit keepers suggest using Bye-Bye Odor.[52] It doesn't cover smells; it eliminates them using a biological, non-toxic formula. Bye-Bye is a good product, but don't try to substitute regular hutch cleaning with it. Use it as an addition to your best practices for a clean and ventilated hutch.

Your Rabbit Must Get Used to Being Handled

Your rabbit will be held and handled by the judge, so it must be prepared for this; otherwise, it might cause a scene by going into a panic.

To prepare your rabbit for being picked up, held, and handled by strangers, follow these guidelines:
- Hold the rabbit gently, supporting its entire body with one hand—stroke the ears, feet, and around its eyes. When your rabbit tolerates this from you, then practice with a family member, and then advance to someone who is a stranger to the rabbit. When the rabbit can tolerate this handling from a gentle and calm stranger, then they are ready to be handled by a judge.[53]

Conditioning Your Rabbit

Show rabbits need to learn to pose for the judges. There may be very specific poses for certain breeds, but if not, these are the general guidelines:[54]
- Align the tip of its front feet with its eyes
- Align the tip of its hind feet with the hip bones
- Make sure the tail is showing

Your rabbit must be trained to sit still in the pose just long enough for the judges to see it and write their score. Posing can be a part of your interaction every day, so it is a normal part of your rabbit's life. As with "all things rabbits," treats are the reward incentive. Gradually increase the time with success. You don't need much time; you just need to be able to pose the rabbit on the ground, step away while the rabbit holds the pose, take a breath, then come back to hold the rabbit. That will do it.

American Rabbit Breeders Association Scorecard
https://arba.net/wp-content/uploads/2021/03/Rabbit-Showman-Scorecard-Updated.pdf

Groom before the Show

Trim your rabbit's toenails at home with pet nail clippers. While holding your rabbit on its back in one arm, gently clip the tip of its nails on each foot. Only trim the white part of the nails and be sure not to cut too close to the pink part. The pink part is called the quick. Cutting into the quick will make them bleed.

If you're nervous about clipping your rabbit's toenails, ask a friend to help you hold your rabbit. If your rabbit's nails aren't clipped for the show, they may snag and fall out or accidentally cut a judge on their hand or arm. Clip its nails the night before or the morning of your show.

Clean any discharge or buildup with a moistened cloth or towel.

Gently clean any light discharge or hay dust around your rabbit's eyes with a damp cloth. If your rabbit has any urine stains or fecal buildup around their rear end, use a different wet cloth to clean that area.

Brush your rabbit before the show to keep its fur neat.

If your rabbit has long hair, it should be brushed every day or two. Short-haired rabbits can be brushed once a week. After brushing, gently rub your rabbit's fur with a damp cloth to remove dead fur. Certain breeds with rex fur, like castor rabbits, should rarely be brushed as brushing can actually damage the texture of their fur.

Showing Your Rabbit

- Keep your rabbit cool and comfortable while transporting it.
  - Traveling in its carrier can cause stress, so add a little hay and a bottle tube waterer to make your rabbit comfortable. Rabbits are susceptible to heat, so keep the air in your car cool to prevent overheating.

Signs of overheating include:
- Hot ears
- Fast and shallow breathing
- Wetness around the nose
- Rapid breathing from an open mouth

If your rabbit is overheating, bring it to a cooler place and dampen its ears with cool water to help get its temperature down.

- Bring a grooming kit, show coop, and your belongings to the show.
- Don't forget to bring grooming supplies, like a brush, scissors, and wet wipes to give your rabbit a quick groom before it is judged. A show coop looks like a wire cage and lets judges handle your rabbit more easily.
- Don't forget your personal items like a phone, wallet, snacks, water, and any paperwork you might need.
- Be present when your rabbit is judged so you can hear feedback.

A show's rules and procedures may vary, but typically, you'll place your rabbit in a show coop when your class is being judged. A judge will evaluate every rabbit and make comments about each one.

It's good "show etiquette" not to reveal which rabbit is yours to other people or the judge. Judges will identify rabbits based on their ear numbers.

Remove your rabbit promptly when the judge is finished.

Rabbits that don't place in the class are removed. Leaving your rabbit on the table too long can be confusing and can slow the judging process. Double-check to make sure it is your rabbit you have removed and not another person's rabbit.

Judging usually begins with all rabbits in a class, then narrows down to the top five remaining rabbits. A judge will call out ear numbers of rabbits that must be removed from the table.

Put your rabbit back in its carrier after it has been judged.

Being around strangers and other rabbits can cause stress, even if your rabbit is well-trained to handle other people. Keep your rabbit comfortable and relaxed by putting it back in its carrier immediately.

When you leave the show, check to make sure you have all your belongings, including your rabbit!

Rabbits used for show can be found for free or may cost up to hundreds of dollars. Remember to check pet shelters for unwanted rabbits. Shelters are usually full of rabbits in the spring.

See these resources for more details.[55] Also, prior to your first show, go to a lot of rabbit shows at a local county fair or other events. Observe and become familiar with the protocols and requirements. You will get a chance to talk to experienced rabbit show people and be able to ask questions. There is nothing in the world like a real, human mentor.

Best Breeds of Show Rabbits for Beginners

Show rabbits require breeds that are both stunning in some way and can be handled and learn to stay calm in a crowd. It is not surprising that there is an overlap between the best pet rabbits and shows rabbits.

The breeds discussed in the pet section are all possible to show, and any of them would be a great breed to start with for showing. There are many other breeds for showing; if you decide to get into fur or meat, you may want to also compete in shows.

In addition to the breeds listed in the pet rabbit section, we will include one more show breed. It was not listed in the beginners' pet section because they require more attention.

Californian Rabbit

Animalcorner.org

- **Weight:** 8–11 lbs.
- **Colors and Fur:** The color of the fur is what makes this breed distinct. The breed standard for these rabbits is very specific. The requirements are described as follows: *"They must have dark markings on their ears, nose, feet, and tail. These markings should be almost black or as dark brown as possible. Their eyes should also be pink, like that of an albino rabbit."*[56] The texture of the fur is courser than the plush, super soft, or satin rabbits.
- **Temperament:** They are well suited for training as show rabbits and make great family pets.
- **Special Considerations:** They will become neurotic and destructive if left in their hutch too much. This is not a rabbit for a person who is away for hours every day.

This is true of many rabbits, but this one is particularly sensitive to being cooped up in the hutch.

- **Lifespan** 5–10 years

Rabbits for 4-H Members

4-H is a wonderful program for children learning how to farm or homestead. Their standards and best practices are rigorous, and they produce adults who are experienced, knowledgeable, and skilled farmers. Even if you don't have children in 4-H, it is worth checking out their resources and protocols for just about anything that has to do with animals, livestock, or growing plants.

Rabbits have a lot of potential for 4-H training, and they are very popular. Being small, they are easier for a child to handle than large livestock, and they can be raised by rural kids who don't have space for larger animals as well. You'll find 4-H kids in rabbit shows whose rabbits are mostly for meat or fur, but also for pet rabbits, breeding for conservation, and agility.

Much of the material we have discussed in this section overlaps with 4-H standards. In the animal and livestock programs, 4-H teaches care and maintenance of animals and breeding, as well as showing livestock at county fairs and other competitive events.

It is worth understanding the 4-H guides for showing; there is great information. Here is a brief outline, but be sure to look for the local 4-H resources.

The rabbitry program for 4-H is for grades 3–12. Below is the link to the national website that includes free downloads of:
- A sample showmanship scorecard
- Rabbit pattern
- Rabbit pattern directions
- You can also purchase the curriculum at 4-H.org[57]

Here is an excellent resource, "Tips for Successful 4-H Rabbit Showmanship."[58]

Conclusion of Rabbits for Show

Showing can be a fun hobby for itself, but it is also an incredible learning opportunity. Even if you just want to have a healthy and beautiful pet rabbit, getting into showing will allow you to meet experienced rabbit keepers. You'll learn a lot, and you don't have to do it forever if you don't want to. I encourage pet rabbit owners to do at least a little bit of showing to deepen your understanding of the best care for your bunny.

Rabbits for Agility Competition and Fun

Homesteaders Rabbit Project[59]

When I first heard of "rabbit agility," I was very suspicious. Rabbits, being prey animals, might bond with owners, but they do not have the same desire to please and obey like dogs have. It seemed to me that there was nothing in their primal desires that would be satisfied by being led around on a leash to jump over obstacles and go through tunnels.

Apparently, I was wrong. Even if it is not the most natural thing for a rabbit to do, the RSPCA states that rabbit agility can be great for the life and health of a rabbit because of the exercise, mental stimulation, and bonding with their human.[60]

You can train your rabbit to compete, or you can just set something up in your house or back yard as part of their mental stimulation and exercise routine. Whatever you do, it is vital that you *train yourself first* to learn how to properly train your rabbit in a way that is safe for their joints and will not make them nervous.

You will see the terms "rabbit agility" and "rabbit hopping." These terms are not synonymous. Rabbit agility is off-leash, and the rabbit goes "around, under, or through" something. Rabbit hopping is also called "show jumping" or "rabbit jumping." Rabbit hopping is an on-leash activity where the owner guides the rabbit over bars or hoops of various heights.[61]

Breeds for Rabbit Agility or Hopping

Almost any breed of rabbit can do some agility at home. Hopping is more specific to breeds that tend to be gifted for it and individual rabbits who take a liking to the challenge. Breeds who are not suited to agility or hopping competition are based on common sense:[62]

- Very large rabbits (the hopping and jumping is too stressful on their joints).
- Rabbits with very large ears (e.g., the English Lop has ears that can get caught in equipment).
- Angora rabbits with super long fur will drag leaves and sticks with them if doing agility outside.
- Hairless rabbits are susceptible to getting cold.

Other than the above criteria to avoid, the best-suited rabbits are medium-sized. They will be stronger and faster than the very small rabbits and more energetic than the large ones. Most agility winners are around 6 to 7 pounds except for the Belgian Hare.

Belgian Hares are rabbits, not hares. They can weigh 6 to 12 pounds, but they are adapted to run fast with long, slender bodies and long, powerful legs. If I were looking for a Belgian Hare for competition, I would try to find one on the lighter end of the scale.

Even though the recommended weight for the best agility rabbit is 6 to 7 pounds, you will find many small rabbits on the agility courses because of their popularity as pets. Don't feel constrained by a smaller size; have fun with your rabbit and join agility activities if you want to.

Equipment for Rabbit Agility

There is equipment you can purchase for agility or hopping courses. If you decide you are serious about agility competition, then it might be advisable to purchase equipment that looks like the courses at the events.

If you are a beginner trying out agility to see if it's for you, or you just want to have fun at home, then making your own course out of free or inexpensive materials might be the way to go.

Let's look at some common agility and hopping components so you can see what kinds of challenges there are; then, we will look at ways you can make them at home before investing in the more expensive equipment.

Rabbit Hopping $135.00[63]

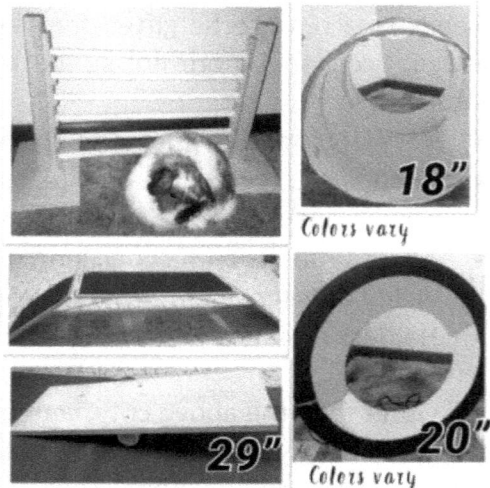

BUNNY HOME GYM

Rabbit Hopping - $185.00[64]

Rabbit Agility – homemade teeter-totter[65]
Weaving poles "Dierentrainer" Channel on YouTube[66]

Before spending more than $300 on equipment, here's a list of the types of activities for agility and hopping:

- Tunnels that go through
- Tunnels that end where the rabbit goes into the end and comes out
- Hoops
- Platforms for the "pause" in the course
- Teeter totters
- Cones or bars for "weaving"
- Bars to jump over as you see in horse steeplechases
- Harness (you don't use neck collars on a rabbit, only harnesses)

The rabbitagility.com site has a lot of great information about equipment, including ideas for making these challenges for free or very little money. Here are links to several pages on their site that have examples of equipment and ideas for making challenges yourself.[67]

One critical note: If you decide to make jumping bars at home, you *must* design them so that if the rabbit's foot touches the bar, it will come off in the direction the rabbit is going. You *cannot* put rope or string or an immovable bar across as the rabbit can get seriously injured if their foot or leg gets caught on the bar. This is one piece of equipment that I would recommend purchasing if you decide you want to get into jumping.

Start with the things you might have at home and see if you and your rabbit enjoy the activity. Then you can decide whether or how to expand your equipment and your agility or hopping adventures.

Training Your Rabbit for Agility or Hopping

Harness and Leash Training

Training a rabbit for *anything* requires time and patience. The training section describes general training for your rabbit's life with you. If you are engaging with agility or hopping, you need specific training.

The general principles of rabbit training all apply to training a rabbit for agility or hopping.
- Never punish a rabbit. Never hit or shout at the rabbit. If you feel impatient or frustrated, then stop the training session.
- Patience and consistency, treats, and positive reinforcement with petting and your voice are crucial.
- Never pull your bunny by the harness.
- *Never* lift the bunny hanging in the harness. This can cause serious injury.

First, you need to teach your rabbit to hop with you on a harness. Be prepared to spend several short training sessions every day. Depending on the rabbit, it may take months for them to be able to adjust and enjoy it.

To begin this, you need to have established a solid level of trust with the rabbit. Don't try harnessing a new rabbit; earn the trust. They must be able to be picked up and handled without struggle or fear and have gained bonding and trust with you.

When this level of trust has been established, then begin to introduce them to a harness.

1) Let the rabbit sniff the harness. Put it in their space (when you are present, so they don't chew it to pieces). If they only want to chew it, then just let them sniff it every day when you are holding them.

2) Gently put it on. To fit the harness, make sure it is tight enough to not come off when a leash will eventually be on it, but not too tight. You should be able to slip one finger between the harness and the rabbit. Depending on their response, let them wear it (with treats) for a few minutes. Gradually increase the time until the rabbit can get the harness put on and spend their outside hutch time happily moving around in the harness.

3) Introduce the leash by clipping it onto the harness and just let the rabbit drag it around. You will need to supervise this to ensure that the rabbit does not catch the leash on something and get stuck—which will cause panic.

4) Start with five minutes and work up to fifteen minutes a day with the leash dragging. When they are used to it, start picking it up and walking with the rabbit. You can call them to walk with you and use treats.

5) Spend a couple of weeks letting them be inside walking with you—after ten days or so, you might feel like your rabbit has progressed to be able to walk with you in a weave or other small challenges, but don't rush it. When you decide to start using the agility equipment, remove all their other normal toys from the area and set up the agility equipment. Don't mix the two—you want your rabbit to understand that when this equipment comes out, this is what they are doing.

6) Once your rabbit is accustomed to walking on the leash with you, then you can take them outside (where there are distractions). If you are training for agility or hopping, you can introduce one or two challenges at a time, train them in those, and then add another.

7) When they are used to some agility or hopping, if you are working up to compete, then you can introduce noise like a television or radio so that they can be a bit prepared for the background noise at an event.

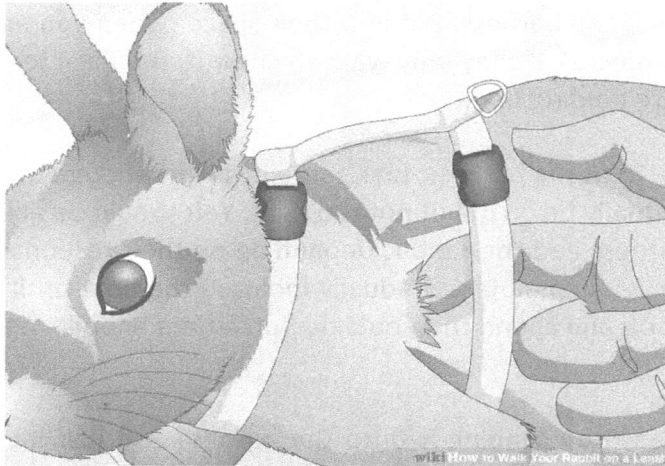

How to Train a Rabbit to Walk on a Leash – WikiHow[68]

Agility and hopping can be fun, and many rabbits love both mental and physical stimulation. This book is for beginners; following the instructions here will give you the basics. If you are advancing to competitive levels in agility, then your rabbit will graduate from going through the challenges with a harness to being off-leash. Here are resources for you.[69] Familiarize yourself with these then seek mentors in your area who can advise and guide you. Find agility events to attend so that you can watch and meet other rabbit owners in your area with the same interest.

Whether for personal fun or competition, enjoy your agility adventures with your rabbit!

Raising Rabbits for Meat and Fur

Families have been raising rabbits for meat for thousands of years. Rabbit meat is delicious and a fine source of protein. It is much easier to raise rabbits for meat than cows, sheep, or other larger livestock.

If you are interested in raising rabbits for meat, then the first thing to do is to check your local ordinances. Many rural areas will allow you a maximum number of rabbits for personal production. Selling rabbits for meat is not in the scope of this beginners' guide.

If you are a homesteader who wants to also use the rabbit pelts, then you will need to consider breeds that are for both meat and fur.

What is described here are the best beginner breeds of rabbits for meat and fur. We will also look at housing that is humane and comfortable for your rabbits. Finally, we will describe humane slaughtering practices and safe processing.

You may be considering rabbits that will give you both meat and fur production. That makes a lot of sense, but if you are a beginner, we recommend you start with this rabbit for meat, and once you've got your systems going throughout a cycle of seasons, then you can advance to meat and fur.

Meat and Fur Breeds

Meat Only

The New Zealand Rabbit – meat only

New Zealand rabbits are the most common meat rabbit. They are easy to care for, and the meat is excellent. There are other meat breeds, but this is the one to start with.
- **Weight:** 10–12 lbs.
- **Colors and Fur:** They are often white but also red, black, brown, blue, and varied.
- **Temperament:** Gentle and docile. Recommended as pets.
- **Special Considerations:** None other than usual rabbit health care.
- **Lifespan:** 5–8 years[70]

Meat and Pelts

Some homesteaders feel that since they are raising rabbits for meat and skinning the rabbit anyway, the pelt should be used. You can also make some money this way. If you have already raised some rabbits for meat and are interested in expanding to meat and fur, then explore these breeds:

The Standard Rex

We met the Mini-Rex in the pet section. The Standard Rex has the same easy personality but is larger and produces both meat and fur. This breed is the most common rabbit raised for fur. We recommend that you start with one of the two following breeds for either meat only or meat and fur. Both are meat rabbits as well.

Champagne d'Argent

This is a beautiful rabbit with a silver color. Its black hairs amongst the gray/white end up giving it a shimmer like silver.

- **Weight:** 9–11 lbs.
- **Colors and Fur:** "Silver" with a soft coat.
- **Temperament:** Good-natured, energetic, and playful. Need exercise. Raised as pets but are not particularly in need of affection which makes them suited to be meat rabbits.
- **Special Considerations:** None, just normal rabbit care.

Satin

The Satin rabbit is an Angora breed, and of that class has the softest and finest fur of all. We don't consider it to be a great beginner's choice because they do take a lot of grooming care.

- **Weight:** 3.5–9.5 lbs.
- **Colors and Fur:** White, grey, brown, or tan.
- **Temperament:** Docile, playful; they need stimulation and ways to play.
- **Special Considerations:** A lot of grooming is required, but if you decide that you want to take on an Angora rabbit specialty, along with raising meat for your family, their pelts are valuable. In terms of health, like all Angoras, their coat makes them more susceptible to the regular rabbit health issues:
    - Wool block – They are more likely to get wool block in their digestive system and end up with a blocked digestive tract.
    - Overheating – Most rabbits are sensitive to heat, but especially Angoras.
    - Diarrhea – They are not more *prone* to diarrhea than other rabbits, but if they do get a bout of diarrhea, you need to carefully clean it off the long, thick fur.

Housing Your Meat Rabbits

See the Outdoor Hutch section of this book for specific examples of hutches that will accommodate meat rabbits. I knew meat rabbit raisers who had Trixie hutches, and they were wonderful. You'll find these pictured in the Hutch section.

Even though your meat rabbits are not pets, they are still breeds who are intelligent and need both physical and mental stimulation and interest for their health.

Provide an enclosure for your rabbits so that they spend at least three hours every day outside their hutch. This is my favorite enclosure; it is usually $250.00—as I'm writing this book, it is listed at $155.00. The shaded spot can move across the roof as the sun moves, which is a big plus.

PawMart 87" enclosure from Walmart $155.00[71]

If you have separation time between your doe and buck, let them alternate having outdoor time. Put some toys and treats and activities in there for them as described in the mental stimulation section. You'll see that cardboard boxes for chewing and exploration go a long way if you don't want to purchase tunnels and toys.

Nutrition and Feeding Practices for Meat Rabbits

Feeding meat rabbits is different from feeding pet rabbits. Your doe is being asked to produce litters of "fryers" (rabbits raised to be butchered for meat) on a regular basis, then nurse them. You want your buck producing the highest quality sperm. Yes, rabbits in the wild are munching on whatever is there, but they don't have human expectations put upon them.

Unlike pet rabbits, we recommend feeding meat rabbits pellets. Do not have them on demand, but measured. Feed them twice a day (morning and evening) at the same time of day.

Most pellets are 16% protein, but you want to feed your doe 18% (you can feed your buck 18% as well). You can add in some hay, but the pellets are carefully designed to offer the right components, including hay.

I love this high-quality rabbit feed for its nutrition and lack of chemicals or additives. Note that one of the designed functions is to "promote growth in larger rabbits."

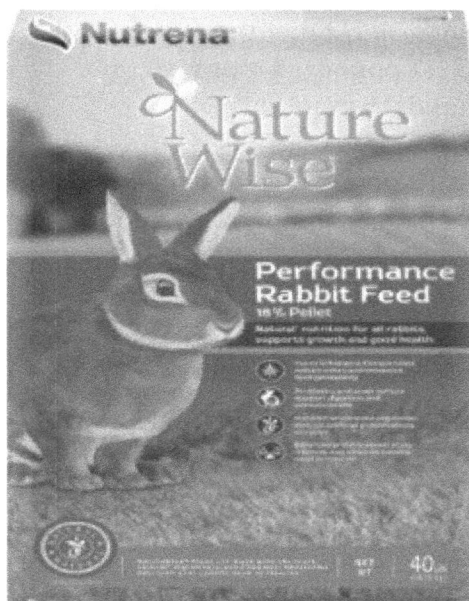

Nutrena World[72]

Breeding Rabbits for Meat

Rabbits are the third most common animals in shelters. They are not valuable to breed casually—people who try often get quickly overwhelmed and end up having to surrender rabbits, or worse, they release them into the wild, which you should never do.

As a beginner's book, we are addressing breeding in this section with the understanding that this is not about raising rabbits for commercial purposes, only for breeding enough for your family.

We assume that you have chosen to acquire a doe and a buck from a reputable breeder for your meat rabbits. Some rabbit farmers say that it's not a problem to breed siblings if you are raising the rabbits for meat,[73] but I have been advised that there can be complications. Make sure they are not siblings.

How Old Should My Rabbits Be for Breeding?

This depends on the size. Don't breed them too young; you have acquired this pair for breeding high-quality litters for you over time. You don't want to overtax them or get them off to a weak start.

Small Rabbits: (under 6 pounds) 4-1/2 months
Medium to Large Rabbits (6–11 pounds) 6 months
"Giant": Breeds (over 11 pounds) 9 months[74]

If you are breeding a 10-pound doe with a 6-pound buck, then wait for her to be ready. Don't breed bucks in a larger category with a doe in a smaller category. It can harm her and cause a miscarriage or stillbirth.

Breeding the Rabbits

If you have some experience with pet rabbits and are expanding into raising rabbits for meat, then you can contact a breeder and offer to purchase a pregnant doe and a buck. The buck will keep her company while she is pregnant but should be removed prior to birth or immediately afterward. Does can get pregnant after birth, and this is not good for their long-term health.

Keep an eye out for your buck and doe's interactions. Always have your animal handling gloves ready if the buck is pushy and the doe isn't receptive; then a serious fight can break out. It's best to keep them in adjacent hutches, so that they can flirt and then be supervised for mating negotiations.

Bucks are usually ready to mate any time. Does can be subtle (to the human eye) about their readiness. One helpful sign is called "chinning." Chinning is when a doe rubs her chin on objects (often on the wire of the wall adjacent to the buck). If she is showing this behavior, then it is a good time to try putting her into the buck's hutch.

Before you put her in with the buck, prepare the space. If you are using cages with wire bottoms, put flooring on the bottom—they can injure their feet on the wire when mating. Use flattened cardboard (they may start chewing the edges, but hopefully, there will be enough floor left by the time they are ready to mate) or wood cut to size.
Don't use anything like linoleum that is slippery or toxic to your rabbits.
Ceramic tile works well if it is a matte finish, so that their feet won't slip.

Put the doe into the buck's hutch.
Be aware of how she is responding.
Have your gloves ready.

If the doe shows signs of not being receptive, separate them. Don't force it; give it time, and try again. Watch for chinning. If they start mating, leave them for about half an hour—they may mate several times, and this is good because it increases the health of the litter.

Remove the buck after they have mated three times or it looks like they are done with two.

Don't be alarmed if the buck falls over. That's totally normal. He'll get up and they will go at it again.

Mark your calendar so you know when the breeding date was.

Is Your Doe Pregnant?

You can check to see if your doe is pregnant at ten days after mating. Gently put the doe on her back on a soft surface such as a carpet or rug placed on a table. I like to use a table because it's more comfortable for *me* to stabilize her and feel for the fetuses.

Gently but firmly push on her abdomen; the fetuses are in the amniotic fluid, so you are feeling them *through* the liquid. "Quality Rabbit Care" offers a great guide for what you're feeling for in their series "Basics of Breeding Rabbits, Part One."[75]

At 10 days, the fetuses are the size of a blueberry.
At 12 days, they are about the size of a marble.
At 14 days, they feel like the size of a large grape or olive.

If you try to feel for pregnancy yourself and get resistance from the doe, or you aren't confident that you can tell whether she is pregnant, make an appointment with a vet tech who can demonstrate the process. If the doe is pregnant, then you can get professional guidance to learn additional care tips.

The first time I did this, the doe was fine on her back (and fortunately trusting and patient), but I wasn't pressing firmly enough. I went to the vet tech, and she demonstrated and helped me find and feel the "blueberries." From then on, I knew what to do.

Strategize Your Rabbit's Breeding Numbers

You want to slaughter your rabbit for meat at about eight weeks. The longer you wait after that, the tougher the meat will be. If you are more interested in the pelts and plan to use the meat for dogs, then you can wait longer. The point is to *think and plan*; otherwise, you will end up with far too many rabbits as they breed exponentially.

How much rabbit meat will you eat? How much room do you have for freezing or canning? Once you have a sense of where you want to start, the best thing to do is to start with a quality pair for breeding. They have a litter, and you separate the buck immediately. A doe *can* breed again at five weeks, but I want my doe to be healthy, happy, and not overbred. I like to let her nourish and tend the litter until eight weeks, when they will be slaughtered.

There's a bit of juggling here—it's not that hard, but you must be intentional.
- You want to have the litters of rabbits in cycles so that you are not overstocked and finding yourself with a bunch of very fertile bunnies.
- When you separate the doe and litter from the buck, bear in mind that when you reunite them, you want to be putting the doe into the buck's hutch, not the other way around. Does are more aggressive about territory. This means that you need two hutches the same size so that there is one for the breeding pair, the buck gets removed after birth to the other hutch, then after the doe is ready to reunite with the buck (and you're ready for another litter), the doe can be put in the buck's hutch, and you'll have enough room for her to tend her litter. This way, they can just go back and forth.

Generally, you can expect litters of 7 to 10. If you decide that 7–10 rabbits every two months is not enough for your family and you want 14–20 instead, then you can think about how to build out this scenario to accommodate two does and maybe two bucks. Fertility in rabbits is "induced" rather than cyclical, so the timing of handling two litters can be done but is tricker. If you have enough space and hutches for separation, you'll be able to rotate your rabbits and control the population.

Pregnancy & Birth

Caring for a Pregnant Doe

The gestation period is about 30 days.

Your doe will need a bit more food. You can talk to your vet about your specific rabbit and any special food they recommend. I just feed my meat doe a little more of her usual pellets. Be careful not to overfeed your doe because it can cause miscarriage.

Don't increase her food all at once—only a little at a time in each meal. To prevent bloat, give her two servings of pellets every day instead of one large meal. We recommended that you do this for your meat rabbits anyway, but during pregnancy, it is a stronger recommendation. Make double sure that she has plenty of fresh water and fiber.

On day 26, prepare a nesting box for her. I use a small cardboard box with unbleached shredded paper that has no ink. If you have a paper shredder, this is easy to provide. If not, you can purchase bags of non-toxic unbleached paper here.[76] Many rabbit owners recommend using hay, but I don't for two reasons:
1) The threat of ear mites in hay.
2) My meat rabbits are used to a diet of pellets. The doe is going to munch on the hay in the new nesting box, and this is a sudden change in diet while pregnant. Not a good combination. If you choose to put timothy hay (a type of grass hay) in your nesting box, be *very* particular about the source.

Your doe will spend time digging in that nest and getting it just right for her babies to be born. Check her every day to see if the nesting material needs to be topped up.

Birth (Kindling) and Afterbirth

Rabbits giving birth is called "kindling." Fortunately, kindling typically doesn't require human intervention. Your rabbit will be nesting for the days prior to giving birth. This is why we recommend giving her a nesting box on day 26 (as described above). Rabbits will usually kindle at night. For the first check, you don't need to pull the nesting box out, just look for movement and listen for squeaks that sound like small birds.

Within twenty-four hours after birth, you'll need to pull out the nesting box, discard the bedding and remove any dead babies. If you have established trust and rapport with your doe, she will probably allow you to touch the babies. This is a huge reason why you want to have a bond with your meat rabbits.

Put the babies back in the nesting box with clean bedding (that has a little bit of unsoiled bedding from the kindling so that the smells make the doe comfortable).

Caring for Baby Rabbits

At birth, rabbits are blind and deaf for the first two to three days. Their only way of receiving communication is through touch. Fortunately, they don't wander; they stay in the nest.

The mother feeds the babies about every twenty-four hours. This is normal; she's not ignoring them. Their bodies are adapted to becoming independent very quickly. If she were in the wild, she would want to decoy predators from the nest. It may look like she's abandoning them, but this is protection.

The babies will scoot on their tummies till about ten days when they start hopping. Be prepared to suffer some serious cute attacks during this time.

The mothers feed the babies for about twenty-one days and then start weaning them. The babies start eating solid food intermixed and finally transition to adult food completely. This happens without teaching or training; the rabbit mothers have a very "hands-off" philosophy of child-rearing.[77]

Weaning will be stressful to the baby rabbits, but taking it incrementally will reduce the amount of stress at any one time. Take about a week with the process. You must sense when is right and observe the mother. Is she trying to wean them? Between five to seven weeks, remove the mother from the nest with the two smallest babies. This is what happens in the wild; the mother just leaves them.

Keeping the two smallest kits with the doe means that she will have an easier transition herself as her milk production will be based on use. You can avoid mastitis this way. Also, those two smallest kits are the smallest because they have been pushed out of the milk bar. This gives them a chance to get some meat on them.[78]

After two days, return the kits to the litter. By twenty-one days, those babies are eating normal food and fully independent. They can stay together as a litter until eight weeks when they are slaughtered. If you are keeping any, then they can be separated and put in separate hutches after nine weeks.[79]

How to Slaughter a Meat Rabbit

By about eight weeks old, your litter of rabbits is ready for slaughter. Slaughtering rabbits humanely is rather easy. You just need the right equipment and supplies set up. We will look at an easy way to set up a station for butchering and processing, then discuss how to do it.

A few points before we begin:
- Have your equipment and supplies set up and at hand.
- Ensure that dogs are inside.
- Slaughter and process your rabbits out of sight of other rabbits.

I learned "hands-on" from a mentor. If you can, that is the best way. The endnotes will include the best resources that explain the process if you must learn yourself.

If you live in a rural area, the easiest way to slaughter a rabbit is with a pellet gun. You put the gun between the ears and aim toward the nose. Rabbit feels nothing and are gone immediately. This is how I learned; it was easy. Having someone demonstrate and then watch me gave me the confidence to do it, and I have never had any mishaps or suffering.

If you are not allowed to fire pellet guns where you are raising your rabbits, then you might consider the "Hopper Popper" which is a popular tool. It's humane, quick, and easy to use.[80] It is a V-shaped tool that is bolted to a tree or post. You put the rabbit's head in it and then pull on the body. It dislocates the cervical spine in an instant, and the rabbit feels no pain.

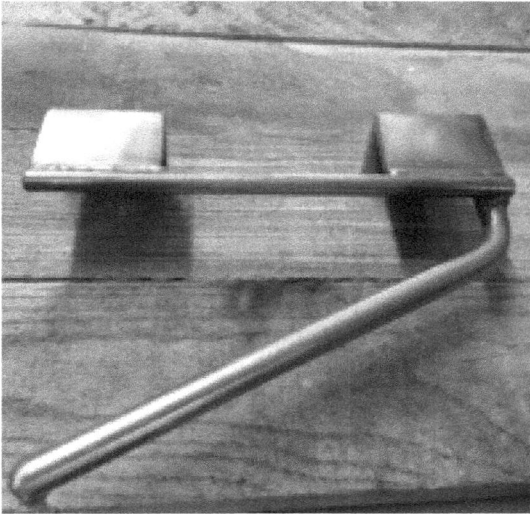

Hopper Popper - $40[81]   The site has instructions for use

You will find conflicting opinions about the Hopper Popper. Some say they use it without any trouble, and they are super happy with it. Others say they have experienced mishaps. You will also find people on forums and blogs who advise that the rabbit be held flat on the ground with a broomstick or metal bar held across the back of the neck, then they say to pull up quickly on the body. This does the same thing as the Hopper Popper (cervical dislocation), but you are having to bend over, hold the broomstick, and perform the action. I've heard of beginners having mishaps with the "broomstick" method, so I highly recommend spending the money for the Hopper Popper to get it at a height that is easy to use and ensures success. The broomstick method can be OK, if your body is comfortable doing it, and you have a mentor to demonstrate and then watch you do it.

Do your research and decide for yourself. This endnote will list a couple of credible blogs that describe the humane slaughter of rabbits and a farmer who offers classes at his farm for butchering and processing rabbits.[82]

How to Process a Rabbit

Once the rabbit is killed, you hang it at your processing station. I like a setup where the slaughter is next to the processing, and I prefer a wall behind the carcass for support. If you don't have a wall available, you can set up a line between trees outside, but then you don't have anything to support you while you're cutting the rabbit.

Wherever you decide to process your rabbit, you will need:

- a place to hang the rabbit by its feet, and
- a tub to catch guts and blood below.
- You will also need two buckets of cold water—one for the skin and one for the meat.

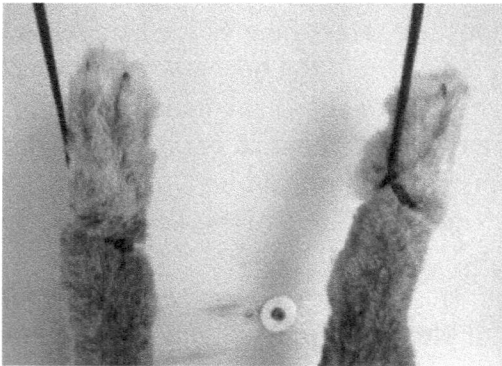

Brick House Acres Rabbitry – They offer classes in butchering and processing.[83]

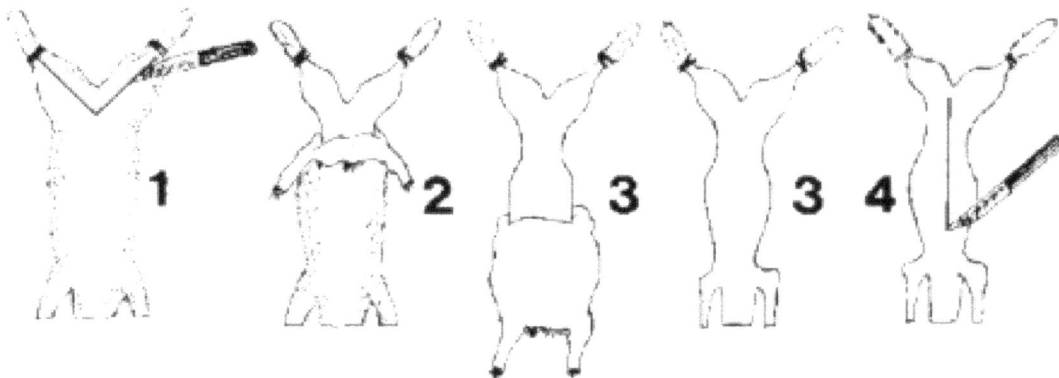

Butch House Acres Rabbitry offers the best illustration I've seen for skinning the rabbit.[84]

I think of skinning the rabbit as taking off a sweater over the rabbit's head. See step 3 in the drawing above. It just peels off. When you get to the head, cut it off from the body where it meets the meat, then turn the skin right side out and cut the head off the skin.

Take the hide and put it in one of the cold-water buckets to keep it fresh and soft while you are processing the meat and cleaning up.

When the rabbit is skinned, you see in step 4 that you cut from the anus down the center of the carcass. Morning Chores suggests that you use your fingers to ensure that the blade does not go through the meat and into the internal organs because you do not want the meat to be contaminated.[85]

After cleaning out the guts and organs, you cut off the feet and put the meat on a table where you can either cut it into pieces or leave it whole. As you are cutting the pieces and cleaning up, have a bucket of cold water ready to put the meat in to keep it fresh till you can get it into the fridge to soak.

Soak the meat for twenty-four hours before freezing or canning. This draws blood out of it. The Morning Chores blog offers descriptions and some images of the processing of rabbits.[86]

Once the meat is soaking in the fridge and you've cleaned up, dump out the water that the hide was soaking in. Refill the bucket with new cold water. Wash the hide thoroughly (but gently) so that all the blood comes out. Use mild soap if you need to, but many rabbit farmers say you don't need this. I've helped my rabbit farmer mentor with this, and they said they have never required any soap.

When your rabbit hide is clean, you need to squeeze the water out. *Do not wring it!* That is like wringing wool—it will stretch it and deform the pelt. Squeeze gently but firmly.

Fill the bucket with *room temperature* water—this time, pay attention to measurements:
- 2 gallons of water
- 1 cup of non-iodized salt
- 1 cup alum

Stir to dissolve.

Put the rabbit pelt in the water (you can process one to six pelts in this solution). Stir it around so that the fur and skin have absorbed and saturated as much of the solution as they can at this moment. Let the hide soak for two days, and stir it twice a day.

After two days, remove the skin from the water (but save the solution). Squeeze it dry, then rinse in *cold* water. You will notice a layer of slimy tissue on the hide. At this point, it can be peeled away. It helps to use a knife, but take care not to cut the pelt.

Add another cup of salt and alum to the bucket of solution you used to soak the pelt. Put the pelt back in it again and stir it around to make sure it's saturated. This time, leave it for a week. As before, stir it twice a day.

After a week, wash the pelt in mild soap and rinse it in lukewarm water. Squeeze (don't wring) the water out of the pelt. Hang the skin up for a couple of hours. It needs to be in the shade, but it can be indoors or outdoors. My rabbit farmer mentor has a standalone clothes rack for this purpose, and she just drapes the pelt over the rack. A shower curtain rack in your house will also work fine.

When the skin is *almost but not completely dry*, take it down to work the hide. Take one small area at a time and massage it between your fingers to make it soft. This is the most work-intensive piece of the process. Working the hide can take several hours, and sometimes you must do it multiple times. If the skin is dry or stiff when it dries, then wet it with a sponge (don't soak it again) and repeat the massaging process. Keep it up until the leather stays soft when it dries.[87]

Raising rabbits for meat and pelts is not for everyone, but it can be an excellent source of protein and some extra cash for a homesteader. The main points to remember as you get started are:

1) House your rabbits humanely so that they are happy and protected from predators.
2) Before you purchase your meat rabbits, think carefully. How much rabbit will you eat? How much freezer or canning space do you have?
3) Confirm the gestation and litters of the rabbits you choose and plan accordingly. You do not want to overbreed. Never release domestic rabbits into the wild.
4) Set up the space and equipment for slaughter and processing.

Rabbit meat is delicious and such an easy way to get meat without large livestock. Bon appetite!

Rabbits as Homestead and Gardening Partners

Rabbit poop and urine provide nutrient-dense fertilizers. Rabbits' feces do not carry the diseases that are concerning for other animals. It composts quickly and can be applied directly to the ground in gardens without composting.

If you are benefiting from the rabbit fertilizer, don't forget the litter! The litter is often hay or paper, so it can be included in a garden or sent to a municipal compost service. Note: if the litter is hay, then beware of the possibility that you'll get alfalfa or grass sprouting in your garden if you apply it directly. Remember that straw is not good for rabbit litter, as rabbits are susceptible to the mites it can contain.

The Rabbit Hutch
Housing a Pet Rabbit in Your Home

There are four basic requirements for housing your rabbit inside:

- A safe, comfy hutch as a bed and resting space.
- Room to roam, get exercise and explore, and be social with you for at least three hours every day.
- The above two spaces need to be near you, not isolated. The ideal is to keep your rabbits in an open plan "great room" where they can be in the "colony" of your family.
- The rabbits need to feel safe from the threat of harm or harassment by children or other pets.

How you provide this in your home depends on the amount of space you have and how it is configured. Some people bunny-proof space, offer nesting places and boxes for space and quiet and just let their rabbits roam free inside that room.

There are some lovely rabbit hutches and "condos" available, but they might give you some "sticker shock." You can find styles that are modern, rustic, or quaint—you'll even see a "Tudor" rabbit hutch below. Here are a few examples of well-made hutches for your rabbits.

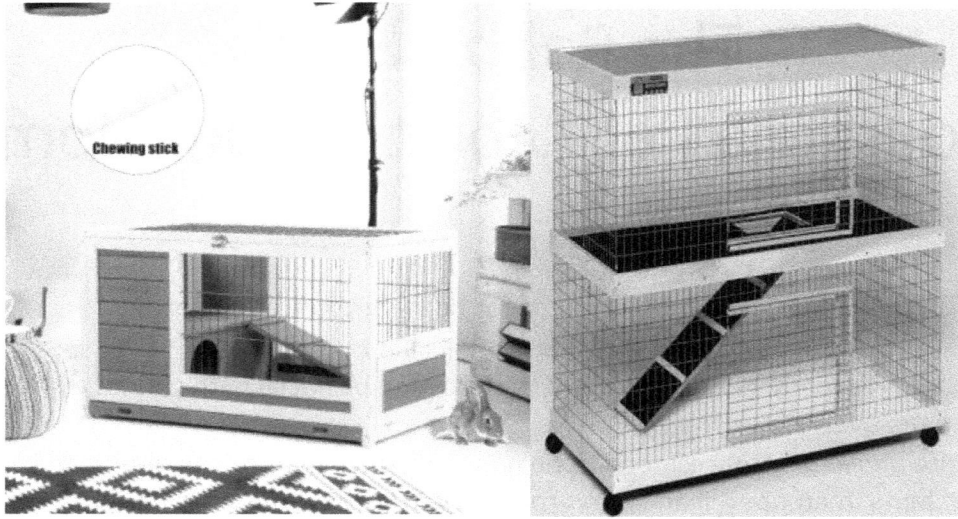

Indoor Rabbit Hutch Wayfair $196.99[88]   Bunny "Condo" at Petwerks $409.00[89]

"Tudor" Rabbit Hutch from Chewy $255.45

There are a lot of DIY resources on the Internet for building your own hutch to keep the cost down.

"Ikeahackers" explains how to make this bunny condo. I love the way this one provides space for use on top.

Ikeahackers "The Bunny Condo"[90]

Another example of a DIY option is from Morning Chores. This is a lovely hutch, and you can get the free plans and instructions to build it.

Morning Chores "50 DIY Rabbit Hutches"[91]

Once you decide on a hutch for your rabbits, you'll need to figure out how to give them an enclosure for their "free time" to play and exercise. Your rabbit will become neurotic if kept in a small space all the time. They need exercise and some interest in their environment. They love to explore a space by sniffing and chewing.

You will want to protect your belongings from getting chewed up. One of the most challenging aspects of bunny-proofing is to ensure that they can't get to wires and charging cables. You need to cover all wires with plastic sleeves or flex tubing or lift them 3–4 feet out of reach of your rabbit. You do not want your rabbits chewing on wires.

Rabbits want to chew on anything they can reach; they love wood and books. Baseboards are often forgotten in rabbit-proofing; they need plastic guards or 2x4s to protect them. Make sure that you don't have poisonous houseplants in your rabbit's reach. You'll need to protect your non-toxic houseplants if you want to preserve them. Don't forget to block access to places like the undersides of beds and items on lower shelving.

When you have protected what you want to preserve, then you can have fun adding in chewing toys and items like cardboard boxes and toilet paper rolls. I love to think about making a bunny play-park, so I'll put a few treats in different places each day, sometimes small piles of vegetable scraps or lettuce greens, and a couple of cardboard chew toys. That way, they are stimulated by needing to explore and find different goodies in different places every day.

It's super fun to use a few cardboard boxes as a "playhouse" and make tunnels and a kind of maze for them with goodies along the way. They also enjoy munching on the boxes as they move along.

Outdoor Rabbit Hutches

If you are keeping meat rabbits or enjoy an outdoor space for your pet rabbits, you may be interested in outdoor hutches.

The outdoor hutch that I strongly recommend is the one below. If you are raising meat rabbits for your family and will have one breeding pair, then get two so that you can separate your buck and doe. You'll be able to move your buck out immediately after kindling and have a place to put him back in that is large enough. It is also good for quarantine purposes if one gets ill. Your pair of meat rabbits have enough space to be happy here.

PetSmart – Trixie Pet Products Outdoor Run Rabbit Hutch $129.00 [92]

If you have spayed/neutered pet rabbits for show or agility, you may want something bigger. Here are two larger options. Both are from the PetSmart website:

Trixie Rabbit Hutch with a View $279.99[93]   Trixie Gabled Roof Rabbit Hutch $399.00[94]
Both are from the PetSmart Website.

Before you invest, remember that you need to have a large outdoor play area as well as a place to separate a rabbit if necessary. You may need to quarantine a rabbit because of illness or injury or separate one because of aggression. Plan for the "what ifs," and it will make your rabbit hobby much easier.

Keeping Your Rabbits Safe

Indoor Predators

Rabbits are prey animals, and they know it. Your cat or dog might not be a danger, but they can also be a stress factor. You may have to separate them, so be prepared with "Plan B" even if you don't think there will be a problem. It's amazing what primal instincts can come out when presented with a rabbit.

Small children or guests who think they can grab your bunny are also "predators" as far as your rabbit is concerned. Teach your children how to gain a rabbit's trust and supervise them. Tell your well-meaning guests how to relate to the rabbit. You must be the protector; otherwise, the rabbit will be stressed, which is both unpleasant and will also erode their physical health.

Outdoor Predators

If you want your indoor rabbit to have time outside, or you have outdoor rabbits who need to get out of their hutches, you need to know who the predators are you are keeping out.

Here are a few common predators in both urban and wild environments:
- Domestic dogs and cats
- Raccoons
- Foxes
- Coyotes
- Aerial predators (hawks, eagles)

These are the most common, but in tropical places, you might have to be aware of large snakes as well.

In rural areas, you will probably have all of those listed above, as well as additional rabbit predators such as:

- Wolves
- Bobcats
- Mountain lions
- Bears being opportunistic

That's a lot of predators. For my pet rabbits (I only have two at a time), I have a small foldable dog pen with a cover to protect from full sun and aerial predators. I like being able to move it so that the rabbits are not peeing on the same spot all the time. I can also set it up in the shade to help keep them cool.

Amazon – New World Pet Products 24" Pen[95]   Amazon – cover for 24" pen[96]
(Note: the cover includes bungee cords.)

If you are raising rabbits for meat or breeding rabbits for other purposes, I recommend a larger pen as you will have one or more litters that need to have space outside the hutch every day. You will also have to rotate your breeding rabbits through the day in order to maintain the separation of those who you don't want to breed at that time. See the meat section for the recommended outdoor enclosure.

Other Supplies and Equipment

Feeders

Indoor Rabbit Feeders

If you have an indoor rabbit, you can give your rabbit constant access to hay. Here are some feeders that work well for indoor rabbit hutches.

Amazon "Rabbit Hay Feeder Bag" $7.99[97]   Amazon – "Mkona Hay Feeder" $12.99 [98]

Here is a feeder that includes a dish for pellets as well.

Amazon – "QSLQYB Rabbit Feeder"[99]

Outdoor Pellet Rabbit Feeders for Meat or Fur Rabbits

These pellet feeders are small, but for meat rabbits, you are portioning twice a day per rabbit.

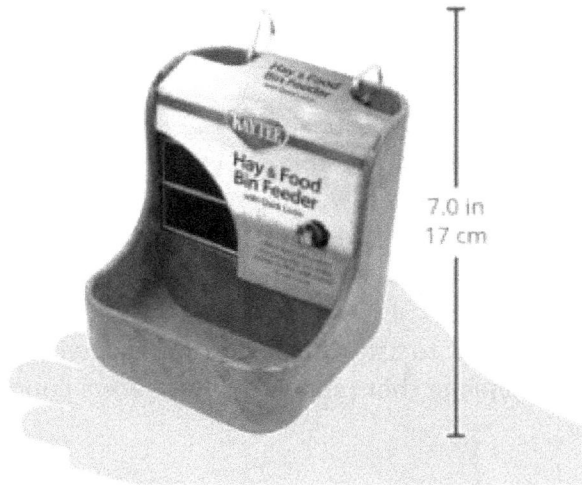

7.0 in
17 cm

Amazon – Kaytee Hay n Food Bin Feeder $24.69 for *two* – ($12.35/ea)

You can also get stainless steel feeders for outdoor homesteading use.

Amazon – "Pet Lodge Steel Small Animal Feeder"[100]
The lid is hinged, so it is easy to put the measured portion of food in at feeding times.

Waterers

Water needs to be fresh, clean, and available on demand 24/7. Nipple water feeders are much better than bowls because they can't be tipped over, and they eliminate mess. They are also more hygienic for your rabbits than having water sit in a bowl that has been licked or stepped in. Normally I pour out and refill my rabbit's waterer when it gets down to half full. The rest is there as a buffer in case I am out longer than expected during a day.

Nipple waterers are built to attach to the outside of wire siding so that you don't have to open the hutch to change them regularly. Here are a couple of examples:

Amazon – "Lixit 64 Wide Mouth All Weather Water Bottle"[101]

If you are housing rabbits outside, you may need to keep your water heated during the winter. This is an excellent choice. I live in Montana, and this has worked for me.

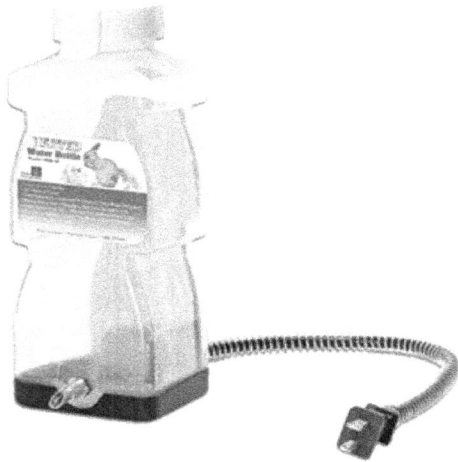

Amazon – "Farm Innovators Model HRB 20 Heated Water Bottle for Rabbits"[102]

The above examples are from Amazon; you may find the same products or high-quality equivalents at your local feed store.

Long (22") Animal Handling Gloves

Have a pair of animal handling or work gloves handy that go up to your elbows. You may need to break up a fight, or you may have an injured or sick rabbit that is frightened and may bite or scratch.

If you are getting pet rabbits who you raise from babies and are docile, tame, trusting, and don't fight, then you can get by with just having the lighter options handy. The problem with that plan is—how do you know they won't fight before you get them? If you are homesteading meat rabbits, I highly recommend having high-quality gloves that are made for animal handling.

The best options for gloves are those made for animal handling like this pair:

Amazon – "RAPICCA Animal Handling Gloves 22" 35.99

They can come in handy if you must remove a snake or other animal from a space on your farm as well.

If you have tame, easy to handle pet rabbits who are not aggressive with each other, at the very least, get something like this:

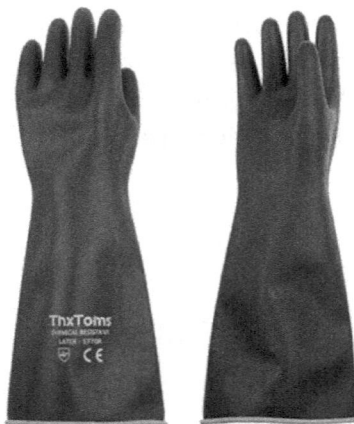

Amazon – "ThxToms Heavy Duty Latex Gloves" $10.99 [103]

You may never use them, but as discussed in the behavior section, even the most docile, tame rabbits can become fierce with their teeth and claws when they fight.

I kept animal handling gloves around for my pet rabbits and was glad I had them when they were sorting out their hierarchy.

Litter Box

Rabbits use a *latrine*, meaning that they like to pee and poop in one place. Most rabbit owners use a cat litter box. I use one in their hutch and one in their roaming area. You can also just have one and move it, but why?

When I acquired indoor rabbits, I invested in another litter box to have in their playtime enclosure. This is very little expense, and although it makes two to clean, it was *less* work because I wasn't moving the litter box in and out of the hutch for their playtime, the cleanup was double but smaller, and they were not concerned with "Where should I poop? Where do I make a latrine?" when they went into their outdoor pen.

Bedding and Litter Filling

Bedding – Does your rabbit need it?

"Bedding" is material placed on the floor of the hutch in addition to the litter box. For indoor rabbits, the problem with putting bedding as well as the litter box in your hutch is you want your rabbit to be clear on where the latrine is. Adding bedding is confusing. If you set up the latrine area as described, your rabbit will have a place to munch on hay while pooping that will make them happy. If you distribute soft material everywhere in the hutch, then your rabbit may not develop the latrine habits that you want.

Bedding can harbor bacteria and set you up for ear mites (hay can carry ear mites, so if it is contained to the feeder, then it is less likely to be transmitted). You can also end up with flystrike (discussed in the health section) and other conditions because of bacteria or parasites harbored in the bedding.

Some people use shredded paper or cardboard, but again it is confusing to the rabbit (where is my latrine?), and the rabbit can have problems from ingesting ink or developing intestinal blockages from eating too much of the paper or cardboard.

The author of "Bunny Lady" suggests ceramic tile. She makes the point that what *we* think a rabbit should have to be comfy (something soft) is not what the *rabbit* necessarily thinks is comfy. Remember, rabbits are heat sensitive. A lot of times, they like to lie on cool tile. Tile is easy to take out and clean, and chilling (not freezing) tiles in excessive heat is also an option. My rabbits have been happy with no added bedding on their hutch floor; if I were to try something, it would be tile as suggested by this author.[104]

Outdoor meat rabbits will need to have bedding during the winter to stay warm. Hay is the recommended bedding for this purpose, be aware that you will have to check and monitor for mold and ear mites during the months that you need insulation.

Litter for the Litter Box

Do *not* use clay-clumping litter or wood shavings (these are harmful to rabbits).[105] The best filler for your litter box is three layers:
- some newspaper on the bottom,
- pellet litter that is rabbit-safe,
- then a little hay on the top.

Use hay rather than straw because straw can harbor the mites that get in rabbit's ears. One rabbit expert points out that straw is stiff, has sharp ends, and can injure your rabbit's eyes.[106] Also, rabbits like to have a pooping space where they can eat and poo at the same time. (Rabbit aesthetics.)

When you change the litter box, leave some of the clean hay and filler that was in the box. Rabbits get stressed when there are changes to their environment, but they are also at respiratory risk if they breathe the ammonia fumes from their urine. I make sure that I leave a little bit of familiar smell to help them feel safe and at home while also ensuring that their litter is clean.

Food and Water

Feeding Your Rabbit

"Best Automatic Feeders for Rabbits"
autoanimalfeeders.com[107]

The main source (70–80%) of your rabbit's daily food should be hay. Baby rabbits just weaned transition to full adult food by starting with alfalfa hay. Adults need to transition to timothy or oat hay. I prefer to rotate these so that they get the widest range of nutrients in their diet.

Use a feeder like the one below. Provide their food "on-demand" so that they have access to it whenever they want. A feeder also keeps the food fresh and clean.

Talk to your vet about nutrition and recommendations for supplementing their hay. I supply pellets and fresh greens. The bag of your pellets will suggest an amount based on their weight, but it's best to discuss this with your vet as you are supplementing hay, not feeding pellets exclusively. The rabbit's need for pellets will decrease as they age, so professional custom advice is best for the long-term health of your rabbit.

The pellets you choose need to be low-protein/high-fiber. These are the pellets that were recommended to me by rabbit mentors as well as my vet:

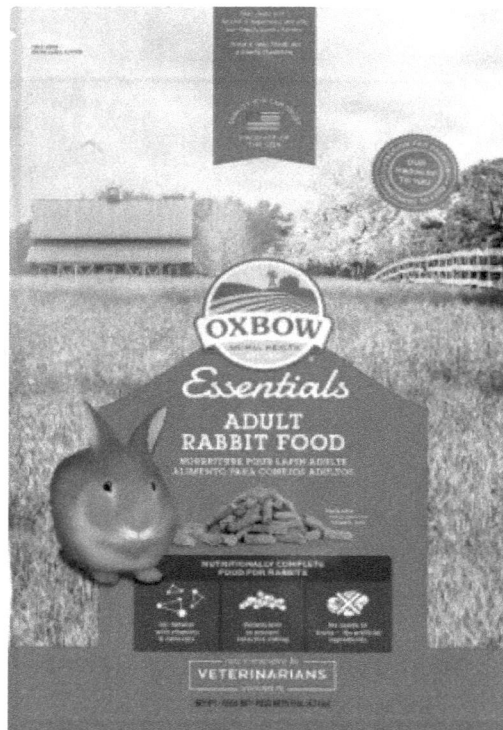

Oxbow Essentials Adult Rabbit Food
Amazon[108]

Vegetable Scraps

Most vegetables and fresh herbs are OK to feed your rabbits, but you should know the ones that are toxic, deadly, or should only be occasional. Some vegetables have components that will make your rabbits sick or die.

Before I got my first rabbit, I thought that I would be feeding it all my green leafy vegetable scraps every day. Fortunately, my vet explained that this is not the case. Getting a solid understanding of which leafy vegetables fall in the "yes," "occasional," or "avoid" category is critical to your rabbit's health.

There are four common reasons that rabbits need to restrict or avoid certain leafy vegetables:

1) High water content
   Vegetables with high water content can give your rabbits diarrhea, gas, or bloating and lead to dehydration. I was not surprised that cabbage falls into this category but discovered that romaine lettuce does as well.

2) Too much calcium
   A veterinarian wrote a great article on this, and it includes a full explanation and a table of the calcium content of various vegetables. In short, too much calcium can *"contribute to sludgy urine and urinary tract problems."*[109]

3) Oxalic acid
   A number of common green leafy vegetables contain high amounts of oxalic acid. Rabbits are not adapted to digest this compound. Oxalic acid can cause problems in a rabbit's urinary tract and kidneys, and cause skin itching.[110]

4) Too much phosphorus
   Rabbits need phosphorus, but too much can cause serious problems, even death. Symptoms of too much phosphorus are *"rash, itching, muscle cramps, numbness, or muscle pain."*[111]

You will see the above four items come up repeatedly in the restricted and toxic lists below, as well as some other reasons that your rabbit must restrict or avoid certain vegetables.

The foods listed in this section are all for *adult* rabbits. See Part Three for the recommended diet for baby rabbits and juveniles.

Whether your bunny is transitioning into adult food, you have acquired a new bunny, or you are introducing a new vegetable to your rabbit, small introductions to new foods are essential. A slow introduction in small amounts applies to the unrestricted foods listed below, as well as the "occasional" foods such as those with medium-high water content. Rabbits are individuals, so just like us, some rabbits' digestive systems will be able to tolerate more than others. Introduce small amounts over time. Observe how your bunny responds.

Let's look at what rabbits can eat without restriction:

Vegetables Your Rabbit Can Eat[112]

- Cucumber
- Zucchini
- Bell Peppers
- Lettuces (*except* iceberg (a toxic substance) and romaine (too much water content) which are to be avoided)
- Arugula
- Watercress
- Endive
- Turnip Greens
- Dandelion Greens
- Chicory
- Raspberry Leaves
- Radicchio
- Basil
- Watercress
- Cilantro
- Parsley
- Wheat Grass

Occasional Vegetables

By "occasional," we mean that your rabbit should only have these once or twice a week:

- Celery: The problem with celery is the strings. They are hard on a rabbit's digestive system. Rabbits do not eat celery in the wild, but they do love it. Split the stalks lengthwise down the middle, then chop them into ½" or smaller pieces. Your bunny's digestive system will be able to handle a bit of this a couple of times a week.[113]

- Spinach: It may seem surprising that spinach might hurt rabbits, but it has high levels of oxalic acid, which is bad for a rabbit's kidney and urinary tract.[114] Spinach has great nutritional benefits, though, so don't avoid it altogether; just include it a little once or twice a week.

- Chard: Treat chard like spinach when feeding your rabbit. It is packed with great nutrients, but contains oxalic acid, so limit its role in your bunny's diet.[115]

- Cauliflower: Rabbits can eat cauliflower florets, leaves, and stems. The reason it needs to be limited is that it can cause bloating and gas.[116] If you cook a cauliflower, you might feed some to your rabbit like this: give a raw floret or two to your rabbit as a treat that day, save the leaves and stem, chop them into pieces, and give small amounts every other day.

- Broccoli leaves and small stems: Like cauliflower, broccoli can cause gas. Expert opinions vary, but we recommended that you only feed your rabbit the leaves and smallest stems occasionally.[117]

- Carrots and carrot tops: Restriction on carrots for rabbits? But "what's up with THAT, doc?!" Carrots are high in sugar and not good for your rabbit as a steady diet. (Root vegetables, in general, are not great rabbit food.) Carrot tops are not much of a concern; experts vary a lot on this—some say carrot tops can be unrestricted, others say that the high calcium can be of concern.[118] If you have a lot of carrot tops, ask your vet about their opinion. Otherwise, keep the carrots down.

- Collard greens: Collard greens also have a lot of calcium and some oxalic acid. That's why it is recommended that your rabbit only eat these occasionally.[119]

- Bok Choy: There is nothing toxic in bok choy, but being a member of the cabbage family, it can cause diarrhea, gas, and bloating. So don't mix it with cauliflower or broccoli, and only give small amounts a couple of times a week.[120]

- Asparagus: Rabbits love asparagus. The great thing about it is that it's low in calcium. The reason it's on the restricted list is that it has a high-water content and can cause diarrhea. Introduce it gradually, and then feed it to them only occasionally.[121]

- Dill: Dill has a lot of calcium, so it needs to be limited. Only about three threads as a topping on other food, and not every day, only twice a week.

- Kale: Rabbits love kale; it's great for a treat or snack. The problem is that it is very high in calcium. It's ok to give a leaf of kale that is about the size of your rabbit's head a couple of times a week.[122]

Fruit

Think of fruits like desserts—occasional treats. Rabbits love fruit and can even form a sugar addiction, rejecting other foods (like a human child). The high concentration of sugar throws off their gut biome, and their bodies do not know how to handle it. Use fruits for treats and maybe dessert once a week. Here are some common fruits that your rabbit will enjoy.

For all but one of these, remove pits or seeds.

- Strawberries: All berries, but my rabbits have always been particularly appreciative of strawberries with their green tops.
- Cherries
- Plums
- Nectarines
- Peach
- Apricot
- Grapes
- Pear without seeds
- Watermelon
- Orange
- Kiwi
- Papaya
- Mango
- Melons (peels and seeds are OK with melons)
- Currants

All of these fruits should be raw, not dried, as the sugar concentration is dried fruit is much higher than fresh.

Vegetables Rabbits Cannot Eat

There are not many foods on this list, but the ones that are here are critical.

- Avocados: Avocado ingestion is likely to be fatal to your rabbit. They contain persin, which protects the avocado from fungus infection. Even a small amount of contact with persin can cause "respiratory issues with fatal consequences."[123] *Remember to protect your rabbit from coming into contact with avocados from children or well-intending visitors as well*.
- Iceberg lettuce: Iceberg lettuce contains a highly toxic component called lactucarium. This substance is sometimes called "rabbit opium." "After the initial jitters, your rabbit will become lethargic, with some diarrhea and GI upsets." Iceberg lettuce also has very high water content.[124]
- Rhubarb: Rhubarb is deadly to rabbits; it is very high in oxalic acid. Symptoms of rhubarb poisoning in rabbits include: lack of appetite, irritation of the mouth, weakness, listlessness, lethargy, diarrhea, dehydration, seizures, neurological damage, death.
- Potato and sweet potato (including leaves): Potatoes and sweet potatoes are starches, so they should not be in your rabbit's diet anyway. They also both contain solanine which is a toxin to rabbits. Solanine causes diarrhea, bloating, and stomach cramps.[125]
- Tomatoes: Tomatoes also contain solanine.
- Onions and all Allium vegetables: Rabbits cannot digest onions or their relatives, and they are poisonous to rabbits, as well. Onions lead to the loss of red blood cells (known as hemolytic anemia) and can also suppress the rabbit's immune system. Sudden death can occur when rabbits eat onions because they can go into anaphylactic shock.[126]

- Garlic: Do not feed garlic to your rabbit for worms or any other reason.[127] At the very least, it will cause digestion issues because it is indigestible to them. Garlic contains high levels of both calcium and phosphorous that are not good for your rabbit.

- Legumes: Lentils, peas, chickpeas, and beans are all legumes, and they are all indigestible to your rabbit. Their digestive systems are not adapted to these foods.

- Seeds and grains: Seeds and grains will not kill your rabbit from toxins, but like legumes and garlic, they are not digestible. They also do not carry the right nutritional components in the right balance for your rabbit.
- Romaine lettuce: Romaine lettuce has very high water content and will give your rabbit diarrhea and possible dehydration. If their leafy greens have a high water content, they will not get the nutrition they need, and over time they can develop kidney and urinary problems. [128]
- Iceberg lettuce.

- Cabbage: Cabbage also has high water content. It is too high for a rabbit for the same reasons as listed above for romaine lettuce.
- Artichoke: Artichokes contain far too much phosphorous for your rabbit and will likely result in the symptoms listed above. Keep your bunny well away from artichokes.[129]
- Root vegetables and starches: With the exception of a very occasional piece of carrot, I do not feed my rabbits root or starchy vegetables. Rabbits are not adapted to dig and root for food; they feed on plants on the top of the ground. We've already discussed that potatoes are toxic because of solanine, but even so, starch is not what their body needs or craves. Their digestive systems do not know how to digest these vegetables very well, so we recommend that you leave them alone.

Fresh Greens

Of course, you can purchase all your rabbit food, but they will appreciate many of your vegetable scraps. You can also cut flower greens such as dandelions from your garden.

Some people choose to take it a step further and grow greens for their rabbits. Since most of their daily diet should be fresh leafy greens, if you have space in a vegetable garden, growing salad greens can be a great way to give your rabbits healthy, fresh nutrition.[130]

Part Three: Maintenance

Health & Happiness

Keep your Rabbit Healthy

Choosing a Veterinarian

Find a vet who is knowledgeable and experienced with rabbits—not all are. Some vets specialize in cats and dogs; others focus on large livestock. Rabbits require their own special needs and have different health challenges. There are vets who have invested in specialty training on rabbits. Look at the educational credentials and background bios of the vets you are considering and find one who can best serve your rabbit.

Vaccinations Required for Fatal Viruses

Vaccine protocols have changed in the rabbit world. There used to be two vaccines required at different times, but now there is a combination vaccine that can be given at five weeks. Talk to your vet about what they recommend for your bunnies. The vaccines cover two fatal and highly contagious illnesses (one of the viruses has two strains):

1) *Myxomatosis*
   Myxomatosis is a viral disease of wild rabbits. It has all but decimated wild rabbit populations in Spain and the UK and even compromised the Liberian Lynx in Spain.[131]
   - Rabbits contract myxomatosis through blood-sucking insects such as mosquitos, fleas, or ticks.
   - It takes up to fourteen days before you will see any symptoms. When you do, here are signs that you need to get your rabbit to a vet immediately. Blue Cross offers this list:[132]
     - *"Swelling, redness and/or ulcers*
     - *Nasal and eye discharge*

- *Blindness caused by inflammation of the eyes*
- *Respiratory problems*
- *Loss of appetite*
- *Lethargy"*
- There is no cure; few rabbits survive if they are unvaccinated. Vaccinated rabbits who contract it might survive with special care.
- Prevent myxomatosis by getting your rabbits vaccinated and keeping up with the recommended boosters. Also, put mosquito-proof wire mesh on your rabbit's hutch, and try to limit your rabbit's exposure to mosquitos. Eliminate any stagnant water you can around your house and keep mosquitos out.
- If you have dogs or cats, make sure you give them flea prevention medication.
- Here are more resources on myxomatosis.[133]

2) Vaccinations also address a second virus that has two strains. "Rabbit (Viral) Haemorrhagic Disease (R(V)HD)"
   - This is another virus that comes from wild rabbits.
   - It is spread through other rabbits, surfaces, or objects in the environment where the rabbit is or through insects.
   - There is no treatment; it is fatal.
   - This virus attacks the internal organs. It is most common that the only symptom of either strain of R(V)HD is sudden death within two days of being infected. If there are symptoms before death, they are:
     - High fever
     - Blood around the nose, mouth, or anus
     - High temperature
     - Lethargy/loss of appetite

Vaccines are highly effective against all three of these viruses. If you take preventative measures, your rabbit should be protected.
- Keep your rabbit away from wild rabbits and any places where they have been. Don't put your rabbit's outdoor pen over the grass where wild rabbits have been.
- Take the same measures as you would with myxomatosis to protect from mosquitos.
- Here are more resources regarding "Rabbit (Viral) Haemorrhagic Disease (R(V)HD)".[134]

Common Ailments

Besides viruses that can be prevented with vaccination and simple attention, there are other ailments that rabbits are prone to. The breeds of rabbits suggested in this book are hardy and not prone to any specific illnesses or conditions. (See this endnote for the one exception.[135]) Rabbits are relatively easy to care for as a pet or livestock animal, but there are conditions that a new rabbit owner needs to know about.

As always, prevention is the best cure. Most of these can be easily prevented with proper care, but it is good to know the signs and symptoms of common ailments.

Ear Mites

Ear mites are tiny. When they first appear, they look like what I describe as "gunk" emerging from inside my rabbit's ears. It is vital to deal with them as soon as you see them because, left unattended, your rabbit will get scabs in the ears, and the mites can spread down the neck and whole body. Lesions will form, which can get infected.

Even with all the best practices of a clean hutch, your rabbits can still get ear mites. The mites can be in the hay you feed them (which is one reason why a feeder is much better than leaving it in a pile on the floor of the hutch or in a bowl). Note that by the time you see the ear mites, they have been there for a while because they start down deep in the ear and move outward.

Early Symptoms of Ear Mites

There may be visual signs of gunk in your rabbit's ear (mites may start in one or both ears). The rabbit will show signs of scratching or shaking of the head. They will probably be doing this before you see any visible signs of the mites.

Remember, the early *symptoms* that you can see indicate the *moderate* stage of infestation in your rabbit. The critters have been busy at work down deep in your bunny's ear canal for 15–20 days before they emerge and start eating the skin of your rabbit's ear where you can see them.

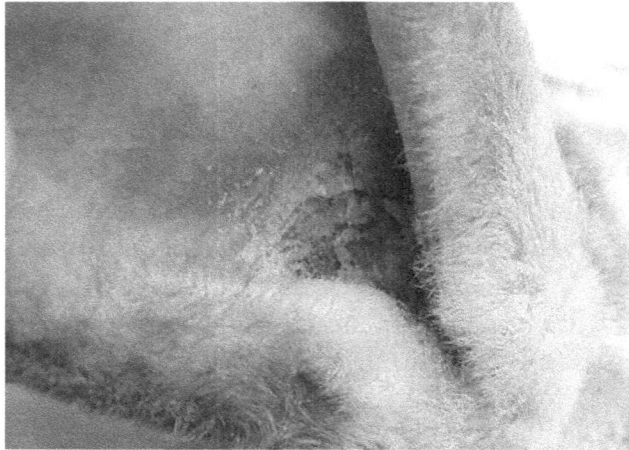

Whitney Living "Ear Mites in Rabbits"[136]

Treatment

Treat the rabbits:

First, you must treat all rabbits in contact with the rabbit you found. Even if you don't see the mites in the other rabbits' ears, they probably have mites deep in their ears that have not shown outwardly yet. The rabbit presenting with a visible infection needs to be quarantined.

Your vet can prescribe an ear wash with Ivermectin or Selamectin. I have depended on this and found it to be effective in stopping those critters and quickly restoring my rabbit's ear health and discomfort. I've never had any infections result from ear mites.

There are homesteaders who say that they've found an effective natural remedy using oil. You can use coconut oil, olive oil, vegetable oil, whatever oil. Add a couple of drops of tea tree oil to it, and you have relief from itching, swelling, and an antibacterial component as well. The oil smothers the mites and they die. Here is a detailed explanation of how to treat ear mites with oil.[137]

Disinfect the space:

You must clean and disinfect the whole hutch, waterer, feeder, and any toys in their living space. Throw away any shredded paper, cardboard, etc. Empty the litter box and disinfect it. Disinfect all combs and brushes.

When the hutch is empty, wipe it down with disinfectant. Ensure that the crevices and corners are cleaned—I use a thin cloth over a table knife after spraying disinfectant on the area.

When you handle your rabbits and disinfect the space, make sure that you wash your hands regularly throughout the process so you don't spread the mites to another rabbit or around your house. PetMD has an excellent article about ear mites.[138]

GI Stasis

Most rabbits should have a diet based on hay. The exception is meat rabbits who are fed pellets that are carefully designed to be a balanced diet. Fiber is essential to rabbits. GI stasis is the blockage of the gastrointestinal tract due to a diet too high in carbs and fats. Sometimes called "the silent killer," GI stasis can cause sudden death.

Symptoms:
Early signs are:
- Your rabbit eating less and less gradually
- Your rabbit isn't drinking as much water as usual
- The stools of your rabbit getting smaller and drier
- Diarrhea followed by small dry stools

If your rabbit is showing these signs, get them to the vet. If you can't, try giving them lots and lots of water and fiber.

The symptoms of the next stages of GI stasis are:
- Your rabbit becomes weak
- Bloating
- Teeth grinding because of pain and discomfort
- Stop pooping completely

This is an urgent situation that can result in death very quickly. PetMD and VCA both offer excellent articles that detail the process and what to expect from veterinary care.[139]

It is easy to prevent GI stasis by providing the proper high fiber diet that your rabbit needs.

Flystrike

Flystrike is common in livestock animals. It is the name for the condition caused by flies laying eggs on an animal's skin. Within twenty-four hours, the maggots hatch under the skin. They eat the flesh of the rabbit and begin to release a toxin that will kill it.

Symptoms:
If you find a maggot on your rabbit or in their hutch, contact your vet. Also watch for flies congregating around the hindquarters of your rabbit (diarrhea is not a symptom but attracts flies). Flies on diarrhea are a precursor to flystrike. Contact your vet immediately if you find a maggot. Clean your rabbit's bottom and check their diet to avoid diarrhea.

In the summer, when there are a lot of flies, watch carefully for flies attracted to your rabbits. Try to limit the fly attractants and accesses in your home. If you've not noticed any maggots, but your rabbit is acting lethargic and not eating like usual, they may have flystrike. See your vet immediately. Do NOT try to wash the rabbit. Get the rabbit to the vet.

Prevention:
Remove fly attractants and access to your rabbits if they are indoors.
Take extra care to ensure that your rabbit has a balanced diet to prevent diarrhea.
During the summer months, when flies are at their peak, watch very carefully for signs of diarrhea in your rabbits. Clean their bottoms and address the possible cause of the diarrhea. If there is diarrhea, clean up the deposits as soon as you possibly can and remove them from the indoor space into an outdoor trash can. Use a sealed plastic bag so you are not attracting flies outside near your home. Clean your rabbit's bottom as soon as possible after defecation if they have diarrhea.

Treatment:
Only a vet can treat flystrike. If you catch it early enough, your rabbit has a good chance of survival. Prevention is the key here. Set up a clean environment and a diet that promotes healthy poop, and you are set up to prevent this terrible, fatal condition. Additional flystrike resources are available.[140]

Coccidiosis

Coccidiosis is a condition caused by a *coccidia,* a parasite that inhabits a rabbit's intestine and ends up infecting the liver. Young rabbits are particularly susceptible to coccidiosis.

Symptoms:

Unfortunately, coccidiosis can be sneaky and subtle. By the time you see symptoms, it has had time to take hold and advance. The symptoms are common to other conditions, so they may indicate coccidiosis or something else. Lack of energy or weakness, lack of appetite, and decrease in water intake.

Call your vet if you see any of these symptoms, but particularly if they are together. If your rabbit becomes dehydrated and has lost weight, then it is possible that the condition has advanced.

Treatment:

Your vet may give you medication to administer to your rabbit at home, or they may feel the need to keep your rabbit hospitalized until they are stable enough to recover at home.

Prevention:

Hygiene is the way to prevent coccidiosis. Keep a clean hutch and litter box. Rabbits eat their own feces so they can infect themselves. Poo left around is more likely to contain the parasites. Water sources can also contain parasites. Do not allow your rabbits to drink from standing water (this is one reason why a nipple feeder is far superior to a bowl). See the MediVet and Vets-Now websites for articles by vets that are informative and helpful to understand coccidiosis.[141]

Heat Stroke

Rabbits are prone to heatstroke. If you live in a place with hot summers, be prepared to bring down the rabbit's body temperature by moving it to a cool area and giving it cool (not iced) fresh water. If you have a fan, your bunny might appreciate that as well.

Another trick I learned in the last couple of years is the use of frozen water bottles or hard icepacks in their hutch. Since I travel and use coolers anyway, I invested in numerous flat Yeti icepacks—the large ones are a perfect size for a rabbit to snuggle up to. If you balance it on a wall of the hutch, your rabbit will lean against it. If you put it in flat, cover it with a thin cloth so their paws don't get damaged, just as you would when you put an ice pack on your skin.

Cabela's[142]

The large Flemish Giants are more prone to heatstroke than the other breeds. This breed should not be raised in a desert region. You will have a clear sign when they are overheated—their ears will turn red.

Symptoms of heatstroke include:
- Shallow breathing
- Drooping ears
- Sluggish movements
- Having a convulsion

Symptoms to Watch for & Troubleshooting

Diarrhea

Diarrhea can be a symptom of a number of serious conditions described above. Something is wrong with your rabbit's tummy, and it needs to be addressed.

Do your best to prevent diarrhea with a balanced diet, but if your rabbit has a bout of it, then clean the rabbit's hindquarters gently as often as possible after pooing and clean up the poo immediately or as soon as you can. Also, find out why your rabbit has diarrhea. If you don't think it is their diet, then contact your vet, who can help you assess.

Snuffly Bunny or Coughing/Sneezing

Rabbits don't get colds. If your rabbit has mucus discharge, then it is likely that there are bacteria causing it. Keeping your hutch clean is the best prevention. Call your vet if your rabbit has snuffles.[143] If your rabbit is coughing or sneezing, this is more urgent. Get that rabbit to the vet; there may be a respiratory infection.

Head Tilt

If your rabbit tilts its head to one side habitually, it may be an ear infection or something else. Check for any other symptoms that are happening (lethargy, snuffles, diarrhea, loss of appetite) and call your vet.

Limping

Rabbits can get injured from being dropped or not held securely. Sometimes they jump from a person's arms. They can get their feet caught on things and get pulls or sprains, and they can be injured in a fight. If your rabbit shows any signs of limping, gently check their paws for injury.

If the paws are injured, then your vet can tell you how to treat it at home or may need to examine your rabbit in person.

If the paws are not injured, and your rabbit is limping, then contact your vet and describe the behavior. Before you call, observe whether it happens all the time or just in certain circumstances.

Rabbit Ears that Droop Suddenly

Know what to expect from the breed of rabbit you have chosen. In some breeds, ears that droop to one side are natural and perfectly normal. The key to watch for is *change*. If your rabbit's ears are not normally drooped and they suddenly change, then it can be a symptom of infection. If the drooping is combined with a loss of appetite, discharge from the ear, or if your rabbit seems irritable, then definitely contact your vet.

If a Rabbit Will Not Hop

If your rabbit stops hopping, observe them and notice whether they are limping or dragging a leg(s). Unwillingness to hop can indicate several conditions:

- Arthritis in the back legs
- Injury or trauma to the hind legs
- Spinal damage
- Splayed legs (a congenital condition)
- A neurological or bacterial sickness, which impacts upon coordination[144]

Contact your vet if your rabbit's movement has shifted from usually hopping to usually walking, especially if you see any limping or dragging of legs.

Hopping or falling from a height (like jumping out of your arms onto the floor) can injure a rabbit. Also, jumping over or off rocks can be too much impact on them. If your rabbits enjoy an enclosure with some space, make sure it is flat without large obstacles.

Sadly, rabbits often get injured indoors. Carpet is the best surface for them; hard floors such as wood, linoleum, or cement will not give them enough traction, so they are skidding regularly. This sets them up for both repetitive stress and injuries by bumping into things because they couldn't stop.

Open-mouth Breathing

Rabbits do not breathe through their mouths naturally. If your rabbit is breathing with an open mouth, then they are having difficulty breathing. Listen to their chest for any wheezing and quarantine them from other rabbits, then call your vet to get advice.

Nasal Passage Irritation

Chemical fragrance, dust, pollen, and other particles can irritate your rabbit's nasal passages. If you see a lot of sneezing and extra mucous, remove the cause of irritation. If left unchecked, it can end up becoming an upper respiratory infection requiring veterinary intervention.

Rabbit Stops Sniffing

This is a sign of bad health. A rabbit should always be sniffing. Sometimes older rabbits will stop sniffing when their health is failing. Call your vet about this for advice.

Conclusion of Health Conditions

Don't be put off by the lists above. If pet owners researched the common health issues for dogs and cats, there would be hardly any pets in homes!

Most of the conditions listed above are preventable. Knowing the symptoms to get an early start on treatment gives your rabbit the best chance for recovery and will mean less effort from you.

Keep Your Rabbit Happy – Mental Stimulation

Besides excellent nutrition, a cozy, clean hutch, and plenty of fresh water, a rabbit needs mental stimulation and exercise.

Your rabbit will become neurotic if kept in a small space all the time. Boredom is unhealthy for most sentient beings. When bored, humans, dogs, cats, goats, horses, pigs, and even chickens will suffer from depression or act out through destructive behavior, self-harm, or aggression. In rabbits, boredom can result in a lack of appetite, destructive chewing, and aggression toward other rabbits and even their humans. They love to explore a space by sniffing and chewing.

Designing Your Bunny Play Park

Giving your rabbit mental stimulation is easy and fun. There are toys you can buy, but you can also make "puzzles" for your rabbit that are cheap or free.

What kind of mental stimulation does my rabbit need?

Your rabbits need to explore, forage for food, play, exercise, and receive affection and grooming from you. Let's look at each need and discuss how to make your rabbits' a playground that is fun for them and easy for you.

Exploring

My favorite way to entertain my indoor pet rabbits has been to give them their greens in their daily "maze" that I make with some cardboard boxes.

- I make paths and a "lounge" for them (larger space to hang out within the maze). I can vary it every time I set the space up.
- If your rabbit shows signs of being anxious in an open space, then make it big enough to toss a toy and get to their food but provide a "roof" so they can relax and not worry about aerial predators. Sometimes I have draped a couple of blankets over chairs to make a 3-sided space that feels safer to the nervous bunny.
- The paths have bits of greens along the way.
- The "lounge" also has balls/toilet paper rolls and food or treat puzzles.

I put my rabbits in the maze three times a day. It takes about two minutes to set up the boxes. I keep them folded flat outside the back door, and then just expand them when I set them up. If I'm using blankets for the lounge, they are handy. Most homes will have all that is needed for a fun and creative bunny playground that is easy to set up and put away.

Rabbits love tunnels. If you want to purchase a tunnel that opens easily and collapses flat, this one is described as "worth every penny."

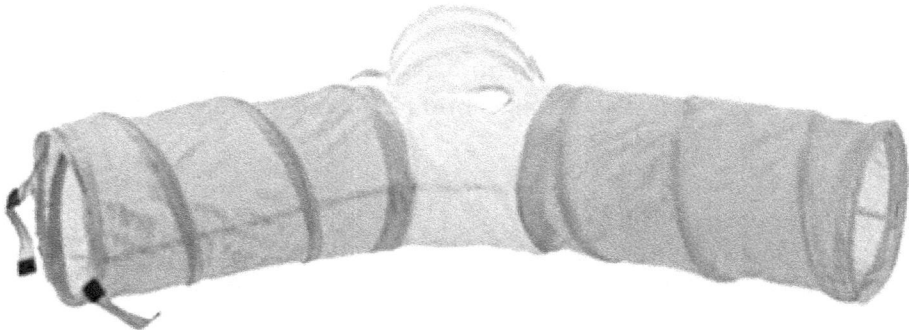

Trixie Play Tunnel for Rabbits
Amazon[145]
(Note: there are two sizes—measure your bunny to get the right one.)

Food-Based Fun

In their natural state, when a rabbit is foraging, they are problem-solving. They need to use their noses to find the best food, hop around to get to it, explore, and nibble.

Years ago, when I had my first house rabbit, I found a blog that introduced me to my favorite way of offering treats. The bulk of greens that I give my rabbit are hung this way in the "rabbit lounge" a few times a day.

"Bunny Logic 101- Rabbits are Smart!"
Bunnyapproved.com[146]

If you want to purchase a food puzzle toy, I like this one the best because it encourages *both* finding the food *and* tossing and playing with the toy carrot.

Kaytee Toss and Learn Carrot Game
Amazon[147]

There are many other options for rabbit "logic" toys and puzzles; my recommendation is that you look for toys with multiple purposes like the one shown above.

It's good to rotate through options so that your rabbits get variety and surprises. These timothy hay balls are a great way for them to get some of their basic feed.

Rabbit Chew Ball
Amazon[148]

Chewing

Chewing on cardboard and toilet paper rolls are great diversions for a rabbit. I also invest in these balls because they sometimes get inspired to play and get exercise when they have the joy of chewing. If you want to purchase chew balls, here are some that serve several purposes of chewing, play, and exercise.

Kathson Rabbit Chew Balls
Amazon [149]

In addition to chewing toilet paper rolls, they will also roll and chase them.

Affection and Grooming

Most rabbits love being petted and groomed. Eventually, they will ask for it. They also love it when people get on the ground and play gently with them. I like to sit in the playground area I've made for my rabbits and roll a ball for them. They will often be happy to be played with, and they will crawl into my lap and just sit to enjoy the petting.

Read the training section for instructions regarding handling your rabbit and gaining trust. We must not force ourselves upon them. After the trust is gained, then your rabbits will feel safe when you approach them. Be aware of your rabbit's communication signals if they do not want to be touched and respect it. See the section on communication for the details of reading your rabbit's body language.

The reason that rabbits are not recommended in houses with small children is that they do not want to be grabbed. Grabbing will likely result in biting or clawing; their primal sense says, "I'm being eaten; I must fight." Rabbits are a good choice for a pet for children around eight—by then, they are usually ready to learn patience and how to negotiate trust.

Conclusion of Mental Stimulation for your Rabbit

Rabbits should not get bored. To keep your rabbits happy, you can make your indoor rabbit playground from items around your house or purchase some toys and tunnels for them. Whatever you do, ensure that your rabbits have mental stimulation, exercise, and your affection to be the happiest they can be.

Grooming

Grooming time is bonding time with your rabbit. It is also your opportunity to look in the ears, check their paws, and notice any changes in their fur or skin that might be of concern. I like to go in stages when I groom my rabbits (or any animal with an undercoat).

First, use a comb to brush out the underfur that is ready to come out. You need to be gentle but firm enough to get the undercoat to come with the comb. If you have not done this before with another animal, you will get the feel of it very quickly. Never tear or pull on the rabbit's fur—this should go smoothly and be a pleasure to your rabbit.

Here is my undercoat tool:

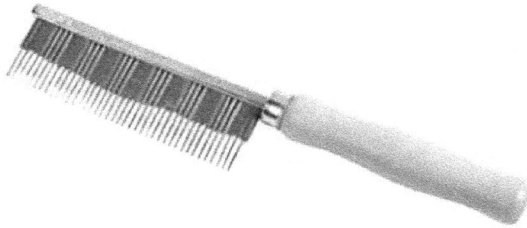

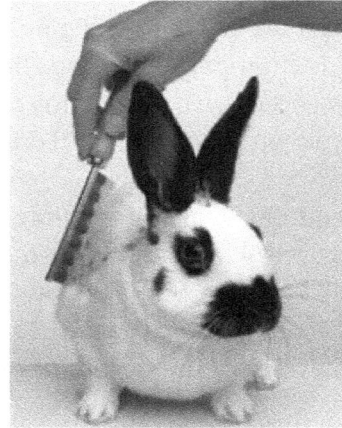

Amazon – Small Pet Select Hair Buster Comb $19.99 [150]

After I've used the comb to get the first level of underfur, then I use another tool to continue. I added this step in 2019 when I watched the video on rabbit grooming on the YouTube Channel 101Rabbits "My Rabbit Grooming Routine." The link is in this endnote.[151]

That video is a great introduction and is very similar to how I like to groom my rabbits. The tool she introduced to me is the Furminator. Yes, it's $30, but she's right— it's worth every penny.

The following image is for the "small pet, short hair" variety. Make sure that you order the right size for your rabbit and the length of their fur. There is something about this tool used after the comb that makes grooming easy and more thorough. My rabbits have loved it.

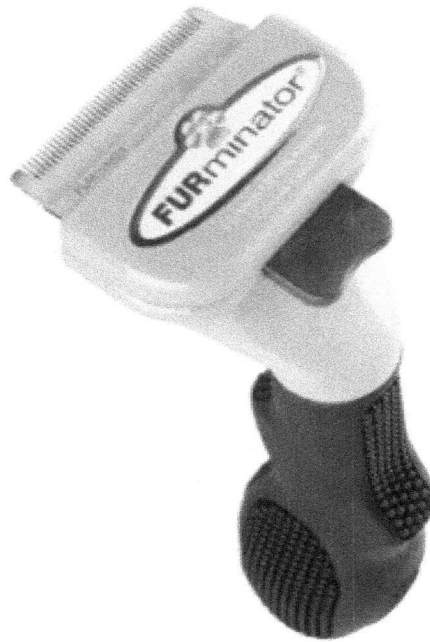

Amazon – Furminator Undercoat Deshedding Tool $30.00[152]

After getting the shedding undercoat, I finish with these silicon brushes on the top. First, I use the green one; then, I finish with the blue one.

Amazon – "Dog Bath Grooming Brush"[153]

I will emphasize that even though it is called a "bath brush," that is *not* how I use it on my rabbit. Never give your rabbit a bath! I like these because they are silicone and very soft on my rabbit. They strap on to my hand for ease and are easy to clean and disinfect.

As I am grooming, I look very carefully for any cuts or issues in the fur. If your rabbit goes outside, it may be important to do a tick check.

Next, I check the ears, eyes, and nose for mites or discharge, respectively. The last thing I do is trim the rabbit's nails. Never use a nail clipper for a human, as they are not shaped correctly for a rabbit's nails. Use a cat nail clipper. Here is an example, but any cat nail clipper will do.

Amazon – Pet Republique Cat Nail Clipper[154]

Like dogs and cats, rabbits have a "quick," which is an area of blood supply in the nail. You don't want to trim into the quick. Trimming the nails outside of the quick is painless, but if you cut into it, there will be bleeding, pain, and likely scratching or biting. Look carefully to find it. If you have another person with you, they might hold a flashlight for you under the nail to be able to see it, or you might take the advice of this video and use a cell phone under the nail to be able to see the quick easier.[155]

Watch the video in this endnote to see how to hold the rabbit and the "2 quick clicks" method. This is *exactly* how I was taught to trim a rabbit's nails over ten years ago (except for the cellphone flashlight trick, of course). [156] The basic idea is that you position the clippers where you think the cut will be and apply a quick "click" of pressure. If the rabbit recoils, then you are probably over the quick and need to pull forward. If not, it is most likely a good spot to trim.

I will say that when I learned, I was not confident. I paid to go to my vet's groomer to watch it done, and then next time, I let her watch me. I clipped the nails of my rabbit twice under supervision before I did it myself at home. There are many more videos that show the exact procedure, so you will probably be able to avoid the cost of the vet. If you feel that you are not confident after watching the videos, then make one appointment to ask the groomer to watch you.

Teeth Checking

Teeth do not need to be trimmed unless they grow incorrectly, or the rabbit doesn't have the diet it needs to keep its teeth healthy.

The only reason that a rabbit's teeth ever need trimming is because of *malocclusion*. Malocclusion is the term for the teeth being misaligned. If this is the case, then get your vet to trim the rabbit's teeth. You may be able to learn but see your vet first and let them guide you with their advice.

Conclusion of Grooming

How often you groom depends on the kind of fur your rabbit has as well as the season. If you own a short-haired rabbit with less undercoat, then once a week will probably work for you throughout the year. If you have a long-haired rabbit or one with a thick undercoat, then you may need to do it once a week in the winter but twice a week the rest of the year. Sometimes indoor rabbits will be out of season in their shedding because, in the autumn, they spend time outside, and then in the winter, they are indoors where it is heated. This will result in winter being the shedding time!

Use common sense. Groom your rabbit regularly so that the fur doesn't build up in the hutch and they aren't ingesting too much of it. Rabbits love to be groomed, and it can be a wonderful time for bonding and trust-building.

Rabbit Behavior

Establishing the Hierarchy in the Colony

Rabbits are hierarchical. There is a clear "pecking order," and they need to be allowed to establish it. There are behaviors that are normal in this process and signal progress or settling; other behaviors are a sign of a problem, and you may need to separate the rabbits to ensure their safety.

When you are first establishing your rabbit colony or when you are introducing a new rabbit, it is best to allow for space outside of the tightly confined hutch. Rabbits are territorial in nature, so it's best to set up a space that is "neutral ground." No one owns this territory as their own, so they can just focus on each other.

Some rabbit owners set up a small, foldable or expandable fence in their back yard—others use a child or dog playpen. Your imagination is the limit on this one, but I would advise that you don't just set them loose in your house so that you can easily keep them in sight and intervene if necessary.

When they first meet, they might lean their body forward with their ears forward and "nose blink." This shows curiosity and getting-to-know-you behavior. It's a good sign.

When a rabbit is relaxed and settled with another rabbit, they might:

- Completely ignore each other. That's totally normal; take it as a good sign.
- Relax together by sitting or lying side by side or by rolling on their back.
- Grooming—if they begin grooming behavior, this signals that the rabbits are settling into the hierarchy.

When you put them in the neutral territory enclosure, start with a short amount of time, then gradually increase the length of time they spend with each other. When you feel like they can hang out together and you don't feel like you must monitor them, then they are ready to be put in a hutch together.[157]

Whether in the hutch or in an enclosure, you may have to deal with rabbits that cannot settle their differences and need to be separated. According to Lou Carter, rabbits have good memories and will hold grudges.[158]

How Rabbits Communicate and How to Understand Their Language

Rabbits use their ears, voices, and bodies to communicate. The following lists are basic *tendencies*. Every rabbit is an individual. One might hold a relaxed position in a way that another holds when they are listening intently. They also combine the ears with sounds and body movements to give more detail and clarity about what it is they are communicating.

Observe your rabbits and get to know what a relaxed (baseline) and happy state looks like. As you watch them, you can discern variances from this baseline state and begin to understand what they mean.

Rabbit Quarrels

If you see grunting, dominance mounting that is resisted, chasing, or fighting, then intervene immediately and separate them for a day—make it two days if there was biting or scratching. Monitor them closely. When you put them back together (in an enclosure, not the hutch), they might just pick up where they left off, or one of them might apologize to the other, and they can reset their status agreements. If they pick up where they left off, then separate them for a longer period. You may find that these two cannot live together, and you must get rid of one or separate them. Some rabbit owners end up with two colonies that they manage separately.

When rabbits who have had conflict make up, they apologize.[159] You'll know this is happening when they touch or rub noses. One bunny will seek to touch noses with the other, and you can watch to see whether the apology is accepted. If the bunny on the receiving end walks away, that is a rejection of the apology, and they need more time separated. If they don't fight in the enclosure, then they can hang out, but they are not ready for the hutch together until this conflict is resolved. If the apology is accepted, both rabbits will stay touching noses and sniffing. Then you'll be able to put them back in the hutch.[160]

Rabbit Ear Movement to Communicate

Rabbits use their ears as communication devices. It makes sense since this is their most versatile and prominent feature. Capable of minute movements, a rabbit's ears will express a lot of detail and tell you how they are feeling. Here are some examples. [161]

Note that if you have a lop-eared rabbit, they still communicate with the ears, but it is more subtle. The movements at the base of the ear are still the same.

- Ears that are almost touching – relaxed
- Up and pointing out – relaxed and happy
- Up and pointing forward – angry
- In a "loaf" – closer together and pointing outward – relaxed and happy
- Ears straight up – listening and sensing to assess the danger
- Ears flat – fear
- Wide apart and flat – fear
- Ears back – your rabbit is angry. If this is in response to something you are doing, stop if possible. The rabbit either wants to be left alone, or they are frustrated with you because they've been trying to tell you something and you have not listened.
- Up and twitching – your rabbit is alert and sensing. This movement serves two purposes—to gather information and to alert others to something that is there.
- One up, one down – "half-listening" – just keeping a check in case something interesting or threatening shows up. I had a rabbit who would do this before every feeding time. She would be lying down with one ear up and would get up and show energy and alertness when she heard the crinkle of the bag with her food.
- Ears lowered all the way, so they touch the rabbit's back – you or another rabbit are about to be attacked. If you don't move away, you'll probably be bitten, clawed, or both. If the other rabbit decides to fight back or responds with resistance, such as grunts, you will certainly have a fight on your hands. If the other rabbit turns its back on the rabbit in attack mode, the attacker may find that acceptable to back off, or it might trigger the fight.

Vocalizations

Besides using their ears, rabbits communicate with sounds and body language. Let's look at common sounds and body movements that often accompany ear communication.

Grunting

Grunting means your bunny is not happy about something. Don't ignore grunting, and if your rabbit grunts with their ears back, then you especially need to listen. You may get scratched or bitten if you don't listen. They might do this when you want to get in to clean their hutch or pick them up. They also grunt at each other when they are telling another rabbit to back off or get out of their territory. If you notice that there are two rabbits who grunt at each other a lot, it is advised that you separate them, then put them in an enclosure outside the hutch to monitor their dynamic.

Snorting

Snorting is less frequent than grunting. Snorting is a milder version of grunting. I think of snorts like a mild "humph," whereas a grunt is a true complaint.

Purring

When a rabbit clicks their teeth together, they are happy. Take it as a compliment if they do this when you hold or pet them. It differs from teeth grinding as it is light and fast. Teeth grinding is a stronger, slower motion that is a sign of pain.

Clucking

Clucking is reserved for expressing joy in food. When your rabbit clucks, note what they are eating because they really love it, and you can mark it as one of their favorite foods or treats. It's their equivalent of "Mmmmm, yum yum yum yum."

Whimper

Like other animals, rabbits whimper from pain or fear. If your rabbit whimpers when you pick them up, they need assurance, calm, and consistency. If they are whimpering in their hutch, check them for injury or any sign of illness. If they are whimpering at another rabbit, then watch the dynamics. If you provide some time together in the enclosure outside the hutch, it might help the whimpering rabbit be less afraid or reveal dynamics that require some intervention.

Honking

Honking is a rabbit's mating cry. It starts softly and then builds louder and louder. If you are breeding rabbits, you'll hear this a lot.

Screaming

A rabbit scream is a death cry. I'm lucky to have only heard it once from a wild rabbit being attacked by a hawk. The hawk was successful. Obviously, you need to find out what is going on if you hear a scream from your rabbit that sounds like a baby or toddler screaming.

Body Language

Standing Up on Back Legs

A rabbit might stand up just to sniff, listen, and watch. They might be curious. Look at their ears and the amount of tension in their body. They also do this to beg. You decide how/whether you want to reinforce this behavior. If they are standing up with the front paws raised like boxing, this is called the "boxing position." This usually happens after a rabbit has signaled that they were feeling defensive, and now they have clicked into all-out aggression mode.[162]

Nudge Nudge...pssssssst!

I love it when rabbits nudge. This seems to be a universal mammal communication to get the attention of the nudged one. Respond to them when they do this; it will help strengthen the bond and connection. Let them know you understand that they want something from you. If they are nudging you repeatedly for something you don't want to do (like give them another treat or continuing to pet them), then break the situation by putting them in or out of their hutch, etc.

Nipping

If you ignore a few nudges, it might develop into a gentle nip. Even if you are not going to do what they want, they need to know they were heard. Nipping can be a sign of displeasure as well. It can be the first step to an escalation that results in an angry bunny. You might get nipped as a warning when you try to clean the hutch or to tell you, "get out of my way." Don't ignore this.

Thumping

Rabbits usually thump as a warning. It works really well in underground burrows, as the sound will travel. You can hear it above ground as well. Thumping can also be a warning because the rabbit is telling you to back off.

Lying Down with Head on the Ground

Again, careful observation of your rabbits is key. This can mean one of two things:

- The rabbit is showing submission to the alpha rabbit
  Or
- The alpha rabbit is telling a subordinate to groom them.

It's important to know the difference.

"I Turn My Back on You!"

When a rabbit turns its back on another rabbit, it can be a rejection of an effort to "kiss and make up" after a fight or a mild expression of annoyance. If a rabbit turns their back on you, they are offended. You need to work to repair the relationship so the rabbit doesn't hold a grudge against you. If the rabbit won't look at you or pulls their ears over their eyes, then it's been a very serious offense.[163]

Binkying

It makes my heart sing when my rabbits binky. Binkying is when they jump and sometimes twist in the air in ecstatic joy and happiness. Like a dog running in donuts and jumping around, your rabbit may be trying to get you or one of their rabbit friends to play. If something triggered this happiness, try to repeat it!

Hopping in a Zigzag

If you see a rabbit zigzagging, they are being chased. The zigzag pattern keeps the predator guessing and spreads the scent around. Find out what is harassing your rabbit and keep them safe.

Training Including Being Handled, and Can a Rabbit be Walked on a Leash?

Rabbits are hierarchical—and they can also be very sensitive to stimulation. Like any animal, if they are bored and don't get enough exercise, they can "act out" in either depression or aggression. Make sure your rabbit is trained to be handled. Unless you are deliberately breeding, spay and neuter your rabbits. This alleviates a lot of behavioral problems. Give your rabbits toys and exercise.

Walking a Rabbit on a Leash

A leash is completely foreign and runs against a rabbit's natural instincts. They can't escape in a situation where they feel exposed. Their natural response is to resist this practice with all their bunny stubbornness. Having said that, you *can* leash-train a rabbit if you are patient, kind, and determined. Eventually, when they realize they are safe, they come to accept and then enjoy it.

If you want to agility-train your rabbit, they must be leash-trained, so clearly, quite a few people are successful in overcoming their rabbit's resistance. See detailed instructions for harness and leash training in the agility section of this book.

Whether or not you intend to participate in agility, the rabbit agility community offers the most comprehensive information about training your rabbit to walk with you on a leash.[164] See the agility section in this book under Choosing a Breed.

Never pick the rabbit up by the harness; this can badly injure them. They can also be injured if they panic when on the leash, so it is critical that you stop leash walking when the rabbit shows discomfort. It does take patience, but some rabbit owners and their bunnies find it rewarding and fun.

Introducing New Rabbits to Your Colony

It is best to purchase two young rabbits who grow up with each other in their hutch. If you acquired only one and are looking for a friend, then take some care introducing them. If you have a choice, it is easier to introduce a female to a male. Females are more likely to be aggressive to a new rabbit in their territory. It's not impossible, though—it just needs a little extra understanding and attention.

Place the new rabbit in a cage next to the established rabbit. Let them sniff and get to know each other. It is best to have two cages, and then you can clean out the hutch so that it becomes "neutral territory" rather than filled with the territorial signals of the established rabbit. Let the rabbits spend time together for a few days in separate cages.

It can take a few days, and you need to be patient and not rush this. You will know they are ready to meet each other when you can see that they like to lie down next to each other where their cages meet.

When you think they are ready, have them meet in a place that is unfamiliar and neutral to the established rabbit. Don't use the living room or kitchen if your first rabbit is used to roaming there at playtime. A bathroom can be a good idea—anywhere that the first rabbit has not been that has a little space for them to meet.

Stay on the floor with them and watch carefully. Have the 22" gloves previously described on and a spray bottle of water on the "jet" setting handy. Some recommend that you have a couple of cardboard boxes with holes cut out as doors so that they can retreat if they get nervous. I've not had the need to do that, but it's a good idea. If all is going well, then the rabbits will be cautious, and then one will begin to establish dominance over the other. This is good and normal.

Sometimes rabbits will immediately fight. This happened once to me. I used the spray bottle, which broke it up and surprised them, grabbed one and got it out, then came back for the other. I had my gloves on, and I was glad that I did. I put them back in their cages and waited another week before I took them to yet another neutral room. They worked it out the second time.

There may be more negotiations to settle the hierarchy, but if that first meeting includes some establishment of dominance, then you will be well on your way.

Conclusion

We have laid out the various functions that rabbits can have and the best breeds for each. After reading this book, you should know whether you want to commit to rabbits, and if so, which ones. Let this book be a resource guide as you gather your equipment and prepare for your new rabbits' homecoming. Use it as a guide for specifics like grooming when you're ready to perform those functions.

With the advice and information in this book, you won't be a beginner for long. Rabbits are one of the most interesting species on the planet. Rabbits can be happy as pets if they have plenty of room to hop, play, and feel safe. The same is true for homestead rabbits that are raised for meat and fur. With a bit of knowledge, a lot of patience, and attention to their needs, raising rabbits and bonding with them is rewarding and fun.

Endnotes

1 https://billingsgazette.com/lifestyles/recreation/wyoming-outdoors-rabbit-hunting-offers-winter-recreation/article_1b28f10f-6ec6-5569-802a-74e828897a55.html

2 https://newrabbitowner.com/wp-content/uploads/2019/10/Rabbit_vision_1080x1080-1024x1024.jpeg

3 https://pyxis.nymag.com/v1/imgs/2af/bc4/45c2d404f0088e8a61e2da0ccdfaaed0cc-ishihara-chart-L0059160.rsquare.w600.jpg

4 A typical daily routine for rabbits - Bunnyhugga.
http://www.bunnyhugga.com/a-to-z/rabbit-behaviour/rabbit-routine.html

5 A typical daily routine for rabbits - Bunnyhugga.
http://www.bunnyhugga.com/a-to-z/rabbit-behaviour/rabbit-routine.html

6 Rabbit Ears: A Structural Look. https://rabbit.org/journal/4-11/ear.html

7 Proprioception is the sense that allows us to maintain balance and sense of orientation in space.
https://medical-dictionary.thefreedictionary.com/proprioception

8 Rabbit Ears: A Structural Look. https://rabbit.org/journal/4-11/ear.htm

9 https://pawtasticpet.com/rabbits/rabbit-ear-anatomy-and-functions/

10 https://www.petplace.com/article/small-mammals/general/rabbit-senses-what-is-it-like-in-their-world/

11 https://www.petplace.com/article/small-mammals/general/rabbit-senses-what-is-it-like-in-their-world/

12 How Good is a Rabbit's Sense of Smell? — Rabbit Care Tips. https://www.rabbitcaretips.com/rabbits-sense-of-smell/

This author notes that while rabbits beat humans and cats in the smell department, they are outdone by dogs who have about 300 million scent receptors.

13 How Good is a Rabbit's Sense of Smell? — Rabbit Care Tips. https://www.rabbitcaretips.com/rabbits-sense-of-smell/

14 How Good is a Rabbit's Sense of Smell? — Rabbit Care Tips. https://www.rabbitcaretips.com/rabbits-sense-of-smell/

[15] Rabbits: Habits, Diet & Other Facts | Live Science. https://www.livescience.com/28162-rabbits.html

[16] https://www.welcomewildlife.com/eastern-cottontail/

[17] https://commons.wikimedia.org/wiki/File:Jack_Rabbit_(14583404194).jpg

[18] http://elelur.com/mammals/snowshoe-hare.html

[19] https://upload.wikimedia.org/wikipedia/commons/e/e3/Arctic_Hare_1.jpg

[20] https://www.fws.gov/sagebrush/images/wildlife/feature/
PygmyRabbit_Seedskadee_Koerner_Feature_Cropped.jpg

[21] https://en.wikipedia.org/wiki/Pygmy_rabbit

[22] https://commons.wikimedia.org/wiki/File:European_Rabbit,_Lake_District,_UK_-_August_2011.jpg

[23] European Rabbit Facts - http://www.earthsendangered.com/profile.asp?gr=M&view=&ID=&sp=11735

[24] Australia is an example of wild rabbits becoming out of balance in their environment and doing a lot of harm.

[25] What Impact Do Rabbits Have on the Environment?
https://bunnylady.com/rabbits-and-the-environment/

[26] For example, Australia has seen incredible destruction. In 1859 European rabbits were introduced but they had no natural predators there so their population became a problem in a short time.

"They began to compete with native animal populations for food and shelter, which also led to the depletion of natural plants and foliage in Australia. Since European rabbits are diggers, they will often dig up and eat new plants as shoots, making it difficult to encourage new growth of Australia's native trees. This also causes soil erosion as the plant life is destroyed."

European rabbits have been released into the wild, in the US, Canada, and South America. They have been a threat to the native wild rabbits in those places.

[27] Ibid.

[28] Ibid.

[29] Data export: Global Rabbit Breeds by Country. DAD-IS (Domestic Animal Diversity Information System). FAO (Food and Agriculture Organization of the United Nations). 21 November 2017. Retrieved 30 March 2018. Referenced by https://en.wikipedia.org/wiki/List_of_rabbit_breeds#cite_note-DAD-IS_2017-1

[30] https://www.pets4-Homes.co.uk/classifieds/2205705-mini-rex-rabbits-herne-bay.html

[31] https://rabbitpedia.com/mini-rex/

[32] https://squeaksandnibbles.com/mini-rex-rabbit/

[33] https://www.pets4-Homes.co.uk/classifieds/2599835-baby-mini-rex-rabbits-caerphilly.html#

[34] https://squeaksandnibbles.com/mini-rex-rabbit/

[35] Mini-Rex characteristics, housing, diet and other information.
https://www.thesprucepets.com/mini-rex-rabbits-5191211

Mini-Rex Rabbits from RabbitPedia https://rabbitpedia.com/mini-rex/

[36] https://animalcorner.org/rabbit-breeds/american-fuzzy-lop-rabbit/

[37] https://aflrc.weebly.com/

[38] https://animalcorner.org/rabbit-breeds/american-fuzzy-lop-rabbit/

https://rabbitpedia.com/american-fuzzy-lop/

https://domesticanimalbreeds.com/american-fuzzy-lop-rabbit-everything-you-need-to-know/

[39] https://www.petguide.com/breeds/rabbit/holland-lop/

[40] https://www.petguide.com/breeds/rabbit/mini-lop/
References photo credit: *Photo credit: Life on White/Bigstock; bobbiesnaps/Bigstock*

[41] https://www.petguide.com/breeds/rabbit/mini-lop/

[42] https://littlefurrypets.com/the-havana-rabbit/

[43] https://pawtasticpet.com/rabbits/havana-rabbit-colors-lifespan-size-and-for-sale/#Temperament_and_behavior

[44] https://pawtasticpet.com/rabbits/havana-rabbit-colors-lifespan-size-and-for-sale/#Temperament_and_behavior

[45] https://pawtasticpet.com/rabbits/havana-rabbit-colors-lifespan-size-and-for-sale/#Temperament_and_behavior

[46] https://www.havanarb.net/

[47] https://amazingpetsforyou.blogspot.com/2017/11/polish-rabbit.html

[48] https://petsnurturing.com/mini-satin-rabbit/

[49] https://www.thezoologicalworld.com/mini-satin-rabbit/#Color_of_mini_satin

[50] Ibid.

[51] Library:Conditioning - RabbitPedia.
http://www.rabway.com/_wiki/index.php/Library:Conditioning

[52] https://www.gundogsupply.com/bye-bye-odor-4oz-concentrate.html

[53] Library:Conditioning - RabbitPedia.
http://www.rabway.com/_wiki/index.php/Library:Conditioning

[54] How to Show Rabbits https://www.wikihow.com/Show-Rabbits

[55] http://www.rabway.com/_wiki/index.php?title=Library:Conditioning

How to Show Rabbits https://www.wikihow.com/Show-Rabbits

[56] https://animalcorner.org/rabbit-breeds/the-californian-rabbit/

[57] https://4-h.org/parents/curriculum/rabbit/

[58] https://qualitycage.com/blogs/quality-rabbit-care/tips-for-successful-4-h-rabbit-showmanship

[59] https://homesteadersrabbitproject.weebly.com/resources/category/selecting-a-good-hopping-agility-rabbit

[60] https://www.rspca.org.uk/adviceandwelfare/pets/rabbits/behaviour/agility

[61] https://rabbitagility.com/

[62] https://www.rabbitagility.com/

[63] https://rabbithopping.com/products/starter-set

[64] https://rabbithopping.com/products/hr-bunny-gym

[65] https://www.rabbitagility.com/?id=62

[66] https://www.youtube.com/watch?v=OMl0obf9B0c

[67] Rabbit Agility Equipment https://www.rabbitagility.com/?id=60

Rabbits at Play https://www.rabbitagility.com/?id=60

Free and Inexpensive Rabbit Agility Equipment
https://www.rabbitagility.com/?id=661

[68] https://www.wikihow.pet/Walk-Your-Rabbit-on-a-Leash

[69] Rabbit Agility Training Articles https://www.rabbitagility.com/?id=58

Training for hopping and other agility specific equipment https://www.rabbitagility.com/?id=61

Agility training your rabbit for obstacle courses
https://smallpetselect.com/rabbit-obstacle-courses-joan-orr/

About harnesses https://www.rabbitagility.com/?id=59

[70] https://www.petguide.com/breeds/rabbit/new-zealand-rabbit/

[71] https://www.walmart.com/ip/PawHut-87-x-41-Outdoor-Metal-Pet-Enclosure-Small-Animal-Playpen-Run-for-Rabbits-Chickens-Cats-Small-Animals-Silver-Green/608691618?wmlspartner=wlpa&selectedSellerId=242&adid=22222222222324807935&wmlspartner=wmtlabs&wl0=e&wl1=s&wl2=c&wl3=74354590368606&wl4=pla-4577954135199737&wl5=&wl6=&wl7=&wl10=Walmart&wl11=Online&wl12=608691618_655802&wl14=outdoor%20dog%20runs%20enclosed&veh=sem&msclkid=1923843e34e01861baefe53b57a1a0dd&gclid=1923843e34e01861baefe53b57a1a0dd&gclsrc=3p.ds

[72] https://www.nutrenaworld.com/product/naturewise-18-performance-rabbit-feed

[73] https://farmingmybackyard.com/can-you-breed-rabbit-siblings-together/

[74] https://qualitycage.com/blogs/quality-rabbit-care/the-basics-of-breeding-rabbits-part-one

[75] https://qualitycage.com/blogs/quality-rabbit-care/the-basics-of-breeding-rabbits-part-one

[76] https://www.walmart.com/ip/Small-Pet-Select-JMWB-Unbleached-White-Paper-Bedding-178-L/605524651?wmlspartner=wlpa&selectedSellerId=17058&adid=22222222222256919497&wmlspartner=wmtlabs&wl0=e&wl1=s&wl2=c&wl3=75247912345827&wl4=pla-4578847476327949&wl5=&wl6=&wl7=&wl10=Walmart&wl11=Online&wl12=605524651_10000017858&wl14=unbleached%20shredded%20white%20paper%20bedding%20rabbits&veh=sem&msclkid=ff269bda72f11e9f4bcbfad3b34630f4

[77] https://farmingmybackyard.com/can-you-breed-rabbit-siblings-together/

[78] https://www.raising-rabbits.com/weaning-rabbits.html

[79] Ibid.

[80] https://theoriginalhopperpopper.com/Hopper_Popper/how-to-guides/

[81] https://theoriginalhopperpopper.com/Hopper_Popper/store/stainless-popper/

[82] https://morningchores.com/how-to-butcher-a-rabbit/

This author offers rabbit butchering and processing classes at his farm. Here he describes the "broomstick method" using a stainless steel bar. https://bharabbitry.weebly.com/humane-rabbit-dispatch.html

[83] https://bharabbitry.weebly.com/humane-rabbit-dispatch.html

[84] https://bharabbitry.weebly.com/humane-rabbit-dispatch.html

[85] https://bharabbitry.weebly.com/humane-rabbit-dispatch.html

[86] https://morningchores.com/how-to-butcher-a-rabbit/

[87] The entire process of tanning is as my mentor taught me. There are a couple of people on the Internet that describe parts of the same process who I also credit here:

Gone Outdoors' description is very close to how my mentor taught me. https://goneoutdoors.com/cure-animal-skins-7358251.html

Morning Chores describes some of what I was taught and offers another way of tanning. https://morningchores.com/tanning-rabbit-hides/

Mother Earth News also discusses tanning rabbit hides https://www.motherearthnews.com/diy/how-to-tan-a-rabbit-hide-zmaz83jfzraw

[88]Wayfair Rabbit Hutch
https://www.wayfair.com/keyword.php?keyword=indoor+bunny+cage&command=dosearch&new_keyword_search=true&class_id=&refid=MX79371125226362.indoor%20bunny%20cage~be&pcrid=79371125226362&device=c&targetid=kwd-79371391996512:loc-4101&channel=BingIntent&msclkid=4db0066c0d3a100df885676857406566&utm_source=bing&utm_medium=cpc&utm_campaign=49_S_1_G_442~WF.S.US.End%20Tables%20LEXICON%201%20(Mobile)&utm_term=indoor%20bunny%20cage&utm_content=442_0_LEX~End%20Tables.indoor%20bunny%20cage.LEX.E

[89]Bunny Condo from Petwerks
http://www.petwerks.com/36-in-double-level-bunny-abode-condo.aspx

[90] https://www.ikeahackers.net/2011/08/the-bunny-condo.html

[91] https://morningchores.com/rabbit-hutch-plans/

[92] https://www.petsmart.com/small-pet/cages-habitats-and-hutches/hutches/trixie-pet-products-outdoor-run-rabbit-hutch-21855.html?cgid=600103

[93] https://www.petsmart.com/small-pet/cages-habitats-and-hutches/hutches/trixie-rabbit-hutch-with-a-view-14658.html?cgid=600103

[94] https://www.petsmart.com/small-pet/cages-habitats-and-hutches/hutches/trixie-gabled-roof-rabbit-hutch-14659.html?cgid=600103#

[95] https://www.amazon.com/New-World-Pet-Products-B552-30/dp/B079P9FDNN/ref=sr_1_29?crid=6DVKT0ZOHIN7&dchild=1&keywords=covered%2Bdog%2Bruns%2Bfor%2Boutside%2Bwith%2Bcover&qid=1627746607&sprefix=covered%2Bdog%2Brun%2Caps%2C339&sr=8-29&th=1

[96] https://www.amazon.com/YGCASE-Universal-Playpen-Security-Exercise/dp/B089RY1SLB/ref=sr_1_25?crid=6DVKT0ZOHIN7&dchild=1&keywords=covered+dog+runs+for+outside+with+cover&qid=1627746607&sprefix=covered+dog+run%2Caps%2C339&sr=8-25

[97] https://www.amazon.com/Rabbit-Feeder-Guinea-Fabric-Storage/dp/B08V8PMQ31/ref=sr_1_2_sspa?dchild=1&keywords=rabbit+feeders&qid=1627832806&sr=8-2-spons&psc=1&spLa=ZW5jcnlwdGVkUXVhbGlmaWVyPUEyTkY4QjkwN1FGTU8xJmVuY3J5cHRlZElkPUEwMDQ0MTMyMTJEQ1VBTENETllMQiZlbmNyeXB0ZWRBZElkPUEwODMwMDM1Mk1YTFNTU0pVSThHMyZ3aWRnZXROYW1lPXNwX2F0ZiZhY3Rpb249Y2xpY2tSZWRpcmVjdCZkb05vdExvZ0NsaWNrPXRydWU=

[98] https://www.amazon.com/Mkono-Feeder-Wasted-Manger-Chinchilla/dp/B072C5LTY5/ref=sxin_14?asc_contentid=amzn1.osa.b70be741-260f-435d-a6a3-1cd77dc4d58a.ATVPDKIKX0DER.en_US&asc_contenttype=article&ascsubtag=amzn1.osa.b70be741-260f-435d-a6a3-1cd77dc4d58a.ATVPDKIKX0DER.en_US&creativeASIN=B072C5LTY5&cv_ct_cx=rabbit+feeder&cv_ct_id=amzn1.osa.b70be741-260f-435d-a6a3-1cd77dc4d58a.ATVPDKIKX0DER.en_US&cv_ct_pg=search&cv_ct_we=asin&cv_ct_wn=osp-single-source-earns-comm&dchild=1&keywords=rabbit+feeder&linkCode=oas&pd_rd_i=B072C5LTY5&pd_rd_r=bfb1c995-2acb-4f09-b420-7d5db8307a12&pd_rd_w=5xIqF&pd_rd_wg=S2BLl&pf_rd_p=8065c57d-81c6-4bce-844a-e686936787b8&pf_rd_r=49GC9KSYAE2XKA60RPYH&qid=1627832905&sr=1-1-64f3a41a-73ca-403a-923c-8152c45485fe&tag=pawtracksdt-20

[99] https://www.amazon.com/QSLQYB-Rabbit-Feeder-Guinea-Chinchilla/dp/B08JGGT4NS/ref=sr_1_10?dchild=1&keywords=rabbit+feeder&qid=1627832905&sr=8-10

100 https://www.amazon.com/Miller-Manufacturing-AF5ML-Rabbit-Feeder/dp/B007Q59DYQ/ref=sxin_14_ac_d_pm?ac_md=2-1-QmV0d2VlbiAkMTAgYW5kICQyMA%3D%3D-ac_d_pm_pm_pm&crid=1XVLJLF1H7FTK&cv_ct_cx=rabbit+pellet+feeder&dchild=1&keywords=rabbit+pellet+feeder&pd_rd_i=B007Q59DYQ&pd_rd_r=2c62735d-9c7e-4d2e-b71d-e0057684fa66&pd_rd_w=VNeEu&pd_rd_wg=gjobu&pf_rd_p=59cd35b3-fd01-41c8-b226-e181f5db7b0f&pf_rd_r=7QX5ADD44ECCWQGV7447&psc=1&qid=1627833877&sprefix=rabbit+pellet+feeder%2Caps%2C353&sr=1-2-22d05c05-1231-4126-b7c4-3e7a9c0027d0

101 https://www.amazon.com/Lixit-All-Weather-Rabbit-Water-Bottle/dp/B0050ICKN2/ref=sr_1_3?dchild=1&keywords=rabbit+waterer&qid=1627835148&sr=8-3

102 https://www.amazon.com/Farm-Innovators-HRB-20-Rabbits-32-Ounce/dp/B000TZ7496/ref=sxin_13_ac_d_pm?ac_md=2-1-QmV0d2VlbiAkMjUgYW5kICQzMA%3D%3D-ac_d_pm_pm_pm&cv_ct_cx=rabbit+waterer&dchild=1&keywords=rabbit+waterers&pd_rd_i=B000TZ7496&pd_rd_r=50863e08-ea9c-459a-9bbe-856a48ca380b&pd_rd_w=ZEW6S&pd_rd_wg=eSnI1&pf_rd_p=59cd35b3-fd01-41c8-b226-e181f5db7b0f&pf_rd_r=1WR84WV2M733KQBH6YAK&psc=1&qid=1627835148&sr=1-2-22d05c05-1231-4126-b7c4-3e7a9c0027d0

103 https://www.amazon.com/ThxToms-Gloves-Resist-Strong-Alkali/dp/B01EAIR38Y/ref=sr_1_19_sspa?dchild=1&keywords=work+gloves+that+go+to+elbow&qid=1627836253&sr=8-19-spons&psc=1&spLa=ZW5jcnlwdGVkUXVhbGlmaWVyPUExVk9XWE1LQ1pLQ1lRJmVuY3J5cHRlZElkPUEwMjc2NzcyM0xVMlhFQlFlMFdkNCZlbmNyeXB0ZWRBZElkPUEwMjU3MjQwU0VTUE1SVTFaSjVQQJndpZGdldE5hbWU9c3BfbXRmJmFjdGlvbj1jbGlja1JlZGlyZWN0JmRvTm90TG9nQ2xpY2s9dHJ1ZQ==

104 https://bunnylady.com/bedding-for-rabbits/

105 http://www.bunnyhugga.com/a-to-z/housing/rabbit-bedding.html

106 https://bunnyadvice.com/bedding-for-use-in-rabbit-hutches/

107 https://autoanimalfeeders.com/best-automatic-feeders-for-rabbits/

https://theoriginalhopperpopper.com/Hopper_Popper/how-to-guides/dispatching-hanging-process/

108 https://www.amazon.com/Oxbow-Animal-Health-Essentials-Food-10Lbs/dp/B0017JANQY/

109 https://www.harcourt-brown.co.uk/articles/free-food-for-rabbits/calcium-and-rabbit-food

110 https://wabbitwiki.com/wiki/Toxic_plants

111 https://whatcanrabbitseat.com/can-rabbits-eat-artichokes/

[112] https://www.oxbowanimalhealth.com/blog/what-are-the-best-vegetables-and-leafy-greens-for-rabbits

https://www.animalwised.com/what-vegetables-are-safe-for-rabbits-to-eat-3518.html

[113] https://squeaksandnibbles.com/can-rabbits-eat-celery/

[114] https://learnaboutpet.com/foods-fatal-to-rabbits/\

https://rabbits.life/can-rabbits-eat-spinach/

[115] https://petcareadvisors.com/rabbits/can-rabbits-eat-swiss-chard/

[116] https://whatcanrabbitseat.com/can-rabbits-eat-cauliflower/

[117] https://www.vetguru.com/can-rabbits-eat-broccoli/

https://www.pfisa.co.za/responsible-pet-ownership

[119] https://animaltattle.com/can-rabbits-eat-collard-greens/

[120] https://rabbitheaven.com/can-rabbits-eat-bok-choy/

[121] https://rabbits.life/asparagus/

[122] https://beyondthetreat.com/can-rabbits-eat-kale/

[123] https://lionheadrabbitcare.com/can-rabbits-eat-fruits/

[124] https://rabbitly.uk/can-rabbits-eat-lettuce/

[125] https://www.animalwised.com/what-vegetables-are-safe-for-rabbits-to-eat-3518.html

[126] *"A severe, sometimes fatal allergic reaction characterized by a sharp drop in blood pressure, urticaria, and breathing difficulties that is caused by exposure to a foreign substance…. The reaction may be fatal."* https://www.dictionary.com/browse/anaphylactic-shock

[127] Some experts suggest garlic as a natural de-wormer for rabbits. This may be a good idea for other animals, but we strongly disagree with its use for rabbits because of the risks and effects. Garlic is not good for rabbits.

[128] https://furrytips.com/can-rabbits-eat-carrots/

[129] https://whatcanrabbitseat.com/can-rabbits-eat-artichokes/

[130] What Impact Do Rabbits Have on the Environment?.
https://bunnylady.com/rabbits-and-the-environment/

[131] https://www.animalwised.com/vaccinations-for-rabbits-225.html

[132] https://www.bluecross.org.uk/pet-advice/myxomatosis

[133] https://2man.org/livestock/myxomatosis-in-rabbits-symptoms-and-treatment-at-home.html

https://www.vets-now.com/pet-care-advice/myxomatosis-symptoms-in-rabbits/

[134] https://www.pdsa.org.uk/taking-care-of-your-pet/pet-health-hub/conditions/rabbit-haemorrhagic-disease-rhd

https://www.bluecross.org.uk/pet-advice/rhd2-rabbit-haemorrhagic-disease-variant-two

[135] The exception mentioned is the Satin rabbit which is an Angora. This rabbit is discussed in the meat and fur section. The long fur of the Satin makes it susceptible to matting from diarrhea, wool blocks from grooming themselves. See the rabbits for meat section to get more details. We do not recommend Angora rabbits for pets for beginners.

[136] https://www.whitneyliving.com/ear-mites-in-rabbits/

[137] http://www.yellowbirchhobbyfarm.com/how-to-treat-ear-mites-in-rabbits-naturally/

[138] https://www.petmd.com/rabbit/conditions/ears/c_rb_ear_mites

[139] https://www.petmd.com/rabbit/conditions/digestive/gastrointestinal-stasis-rabbits-it-really-hairball-causing-blockage

https://vcahospitals.com/know-your-pet/gastrointestinal-stasis-in-rabbits

[140] https://www.medivet.co.uk/pet-care/pet-advice/pet-summer-safety/flystrike-in-rabbits/

https://www.vets-now.com/pet-care-advice/flystrike-in-rabbits/

[141] Ibid.

[142] https://www.cabelas.com/shop/en/yeti-ice

[143] https://morningchores.com/rabbit-diseases/

[144] Can Rabbits Walk or Just Hop? —
https://www.rabbitcaretips.com/can-rabbits-walk-or-just-hop/

[145] https://www.amazon.com/Trixie-62791-Play-Tunnel-Rabbits/dp/B001SI08U0/ref=as_li_ss_tl?ie=UTF8&linkCode=sl1&tag=petsmentor0-20&linkId=7392e8b1c9f0df29ca876ed67effa812&language=en_US

[146] https://bunnyapproved.com/bunny-logic-101/

[147] https://www.amazon.com/Kaytee-Toss-Learn-Carrot-Game/dp/B06XSGSWMS?tag=smallpetjournal-20

[148]
https://www.amazon.com/gp/product/B08341TYXV?pf_rd_r=9Y1G38BW3X12A3FWR1V8&pf_rd_p=5ae2c7f8-e0c6-4f35-9071-dc3240e894a8&pd_rd_r=a97ab3aa-38c7-4daa-84a0-5fc516025465&pd_rd_w=IOnOH&pd_rd_wg=XwaGb&ref_=pd_gw_unk

[149] https://www.amazon.com/kathson-Rabbits-Grinding-Activity-Chinchilla/dp/B08P8JZ5KL/ref=sr_1_19?dchild=1&keywords=rabbit+toy+balls&qid=1626456886&s=pet-supplies&sr=1-19

[150] https://www.amazon.com/Small-Pet-Select-HairBuster-Comb/dp/B06ZZXF81G/ref=sr_1_1_sspa?dchild=1&keywords=hair+buster+comb&qid=1627934419&sr=8-1-spons&psc=1&spLa=ZW5jcnlwdGVkUXVhbGlmaWVyPUExU0ZZSUdFSDFaNU81JmVuY3J5cHRlZElkPUEwNTE1ODkyM1U1MzY2VkhEV0QySCZlbmNyeXB0ZWRBZElkPUEwNTY1MDQ3OVA4NUFKWDY2RlBKJndpZGdldE5hbWU9c3BfYXRmJmFjdGlvbj1jbGlja1JlZGlyZWN0JmRvTm90TG9nQ2xpY2s9dHJ1ZQ==

[151] https://www.youtube.com/watch?v=hL78-fmflWE

[152] https://www.youtube.com/watch?v=hL78-fmflWE

[153] https://www.amazon.com/Dog-Bath-Grooming-Brush-Scrubber/dp/B09172PFP1/ref=sr_1_4_sspa?dchild=1&keywords=small+pet+rubber+brush&qid=1627934658&sr=8-4-spons&psc=1&spLa=ZW5jcnlwdGVkUXVhbGlmaWVyPUExVFQ2WElYWFdFQTU3JmVuY3J5cHRlZElkPUEwMjc0NDcwTFhFTjNNRDhPMVEyJmVuY3J5cHRlZEFkSWQ9QTAwNjI1NjUxNVBIWE84VkNNMkQ2Q2JndpZGdldE5hbWU9c3BfYXRmJmFjdGlvbj1jbGlja1JlZGlyZWN0JmRvTm90TG9nQ2xpY2s9dHJ1ZQ==

[154] https://www.amazon.com/Pet-Republique-Professional-Nail-Clippers/dp/B01GBSSKVU/ref=sr_1_2_sspa?crid=2LYC8TPK14Y3J&dchild=1&keywords=cat+nail+clipper&qid=1627937066&sprefix=cat+nail+clipper%2Caps%2C304&sr=8-2-spons&psc=1&spLa=ZW5jcnlwdGVkUXVhbGlmaWVyPUFFVFFYQ1ZGVENYSTYmZW5jcnlwdGVkSWQ9QTAzNjAwMDEyMkNWTE5GNGUcxTzRaJmVuY3J5cHRlZEFkSWQ9QTAzNzQ1OTA5TFM2MExSRzBCR0ggmd2lkZ2V0TmFtZT1zcF9hdGYmYWN0aW9uPWNsaWNrUmVkaXJlY3QmZG9Ob3RMb2dDbGljaz10cnVl

[155] YouTube 101 Rabbits "How to Trim a Rabbit's Nails" https://www.youtube.com/watch?v=rvausPZOJ5A

[156] YouTube – 101 Rabbits "How to Trim a Rabbit's Nails"
https://www.youtube.com/watch?v=rvausPZOJ5A

[157] How Do Rabbits Communicate with Each Other?
https://www.rabbitcaretips.com/how-do-rabbits-communicate-with-each-other/

[158] How Do Rabbits Communicate with Each Other? — Rabbit Care Tips.
https://www.rabbitcaretips.com/how-do-rabbits-communicate-with-each-other/

[159] How Do Rabbits Communicate with Each Other? –
https://www.rabbitcaretips.com/how-do-rabbits-communicate-with-each-other/

[160] Ibid.

[161] This list is based upon the excellent article by Lou Carter, What Do Rabbits Ear Positions Mean?
https://www.rabbitcaretips.com/rabbits-ear-position-meaning/

[162] Ibid.

[163] Ibid.

[164] https://rabbitagility.org/about/

http://www.mncompanionrabbit.org/rabbit-agility/

https://be.chewy.com/how-leash-train-rabbit/

www.ingramcontent.com/pod-product-compliance
Lightning Source LLC
Chambersburg PA
CBHW081240220326
41597CB00023BA/4322